T0138943

Additive Manufacturing

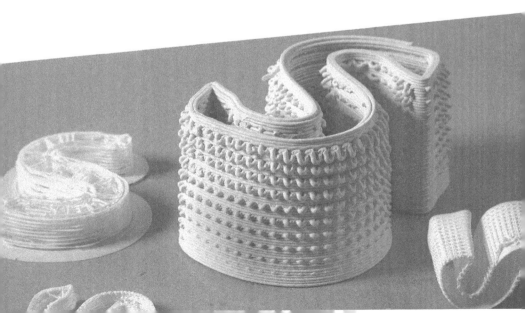

Additive Manufacturing

Design, Methods, and Processes

edited by

Steinar Killi

PAN STANFORD PUBLISHING

Published by

Pan Stanford Publishing Pte. Ltd.
Penthouse Level, Suntec Tower 3
8 Temasek Boulevard
Singapore 038988

Email: editorial@panstanford.com
Web: www.panstanford.com

British Library Cataloguing-in-Publication Data
A catalogue record for this book is available from the British Library.

Additive Manufacturing: Design, Methods, and Processes

Cover art by William Lavatelli Kempton
Editor's photograph by Thor Hestnes

ISBN 978-981-4774-16-1 (Hardcover)
ISBN 978-1-315-19658-9 (eBook)

Printed in the USA

Contents

Preface   xi

1. **Scope of the Book**   **1**
   Steinar Killi

   1.1 The Magic of 3D Printing   1
   1.2 Legal Issues   4
   1.3 The Power of Rhetoric   6
   1.4 Maturing of Technology   7
   1.5 "We Are All Designers"   8
   1.6 From Rapid Prototyping to 3D Printing   13
   1.7 The Chapters of the Book   18

2. **A Design Sociotechnical Making of 3D Printing**   **21**
   William Lavatelli Kempton

   2.1 Introduction   21
        2.1.1 Disciplinary Boundaries and Claims
               to 3D Printing   21
        2.1.2 Introducing a Sociotechnical
               Perspective to 3D Printing   22
               2.1.2.1 Sociotechnical development
                        from a design perspective   23
        2.1.3 Outline   24
   2.2 Socially Constructed Technologies and 3D
        Printing   25
        2.2.1 The Relevance of Social Groups   25
        2.2.2 From Video Production to Material
               Production   26
        2.2.3 Technologies for Additive Making   26
        2.2.4 Critical Theories and Studies of
               Technology   27
        2.2.5 Unpacking the Views of 3D Printing   28
        2.2.6 Socially Constructed Perspectives of
               Additive Making   29

		2.2.7	Relevant Social Groups as Part of a Technological Frame	30
2.3		The 3D Printer Inventors		31
	2.3.1	The First Wave of 3D Printer Inventors		32
	2.3.2	The Second Wave of 3D Printer Inventors		34
2.4		Business Perspective of 3D Printing		36
	2.4.1	Yet Another Industrial Revolution		36
	2.4.2	Toward Economies-of-One		37
2.5		Designers' Perspectives of 3D Printing Futures		40
	2.5.1	Design and Additive Manufacturing		41
	2.5.2	Designing with Technology		42
	2.5.3	An Undetermined View of Design		43
2.6		A Layperson's Perspective of 3D Printing Futures		44
	2.6.1	A Layperson as a Maker		45
	2.6.2	Making in a Learning Environment		46
2.7		Discussions and Conclusions		48
	2.7.1	Summarizing the Perspectives		48
	2.7.2	3D Printing Futures		48
	2.7.3	Constructing a View of Sociotechnical Development		49
		Appendix: Technologies for 3D Printing		50

3. **AICE: An Approach to Designing for Additive Manufacturing** — **75**

Steinar Killi

3.1		AICE: An Operational Model		81
	3.1.1	Adapt		81
		3.1.1.1	Design thinking	81
		3.1.1.2	Multitypes	84
		3.1.1.3	Models describing a typical design process	85
		3.1.1.4	Methods used during a design process	89
	3.1.2	Integrate		98
	3.1.3	Compensate		100
		3.1.3.1	Spare parts	102
		3.1.3.2	Production aids	103
		3.1.3.3	Enhancing the design	103
	3.1.4	Elongate		106
3.2		Using the AICE Model and How the Drinking Container Came Out		109

4. **The Impact of Making: Investigating the Role of the 3D Printer in Design Prototyping**   **111**

    William Lavatelli Kempton

    4.1   Introduction   111
            4.1.1   Prototyping as Design Development   111
            4.1.2   Making as a Critical Practice   112
            4.1.3   Outline   113
            4.1.4   Methods   114
    4.2   Background   115
            4.2.1   From Rapid Prototyping to Additive
                    Manufacturing   115
            4.2.2   Ubiquity and Stratification of 3D
                    Printing   117
            4.2.3   Contexts for Additive Making   118
            4.2.4   Hybrid Artifacts   120
            4.2.5   Making Representations as a Way
                    of Designing   121
    4.3   Prototypes and Design Representations   121
    4.4   The Changing Character of Design   123
            4.4.1   New Product Development   123
    4.5   Situating AM Prototypes within Design
            Practice   125
            4.5.1   Developmental Prototypes   125
            4.5.2   Initial Concept and Maturation of
                    the SunBell Lamp   126
    4.6   Design Representations and Multitypes
            in Product Design   129
            4.6.1   Multitypes in Rapid Prototyping   131
    4.7   Multityping in Additive Manufacturing   134
            4.7.1   Popular yet Professional?   134
            4.7.2   Integrating AM in Product Design   135
            4.7.3   Toward the Releasetype   136
    4.8   Conclusions   137

5. **Visual 3D Form in the Context of Additive Manufacturing**   **143**

    Nina Bjørnstad and Andrew Morrison

    5.1   Introduction   143
    5.2   Aesthetics   144
    5.3   Design, Action, and Profession   147

5.4	Ideals and Origin	148
5.5	Relevance of the Historic Model for Tomorrow's Form Givers	149
5.6	The Evolution of Form Model	150
	5.6.1 Why Clay?	151
	5.6.2 A Close-Up on Form	153
	5.6.3 Distorted Forms	155
	5.6.4 Intersectional Forms	157
5.7	Familiarity	160

6. Potential of Additive-Manufactured Products in Building Brands — **165**

Monika Hestad and Viktor Hiort af Ornäs

6.1	Product Role in Brand Building	168
	6.1.1 Role of Design Elements in Building a Brand	168
	6.1.2 Brand Story and Product Story	169
6.2	Additive Manufacturing as One of Many Other Drivers That Affect a Product's Role in Building Brands	171
	6.2.1 Actual and Intended User Experience	171
	6.2.2 Internal Drivers	173
	6.2.3 External Drivers	174
6.3	How Additive Manufacturing Is Used in Building a Brand	175
	6.3.1 Mykita	177
	6.3.1.1 How AM is used in the products	179
	6.3.1.2 Design elements	179
	6.3.1.3 User experience	179
	6.3.1.4 Drivers	180
	6.3.1.5 The Mykita brand story	182
	6.3.2 pq Eyewear by Ron Arad	184
	6.3.2.1 How AM is used in the products and in branding	186
	6.3.2.2 Design elements	186
	6.3.2.3 User experience	188
	6.3.2.4 Drivers	189
	6.3.2.5 The pq eyewear brand story	190

6.4 Potential of Additive-Manufactured Products
in Building a Brand 191
  6.4.1 How They Used AM in Building a Brand 191
  6.4.2 Opportunities in New Production
Techniques 193
  6.4.3 Form Freedom and Brand Development 194
  6.4.4 Potential of Disruptive Stories 195

**7. A Tale of an Axe, a Spade, and a Walnut:
Investigating Additive Manufacturing and Design
Futures** **199**

Andrew Morrison

7.1 Prelude 199
7.2 Queries 201
  7.2.1 On Discursive Design 203
7.3 "Problems" 205
  7.3.1 Design, Narrative, Futures 205
7.4 Essayistic 206
  7.4.1 Narrative 208
7.5 Promotion 210
  7.5.1 Intersections 213
7.6 Foresight 214
  7.6.1 Scenarios and Futures 216
  7.6.2 The Fictive and Nondeterminist
Futures 217
7.7 Reflections 218
  7.7.1 Toward the Additive in Discursive
Design 220
  7.7.2 Design Baroque Futures 220
7.8 Generative Visions 222

Index 231

Preface

Just before the year's end in 1997, the Oslo School of Architecture and Design (AHO) bought its first rapid prototyping machine. This was an early adaptation to an emerging technology, and we saw its potential for experimentation and pedagogy relating to product design. The machine used was Sinterstation 2000, a powder-based system that, up to this date, had been used by Chrysler. Over the next 20 years, AHO purchased several new machines, covering almost all available technologies relating to additive manufacturing, as this field has matured from rapid prototyping through the more recent commercialization and marketing of the now popular and "maker"-oriented 3D printing.

The terms describing this technology have changed over the period I have been exploring it, and this reflects largely a focus on technology. The terms have shifted somewhat from "rapid prototyping" to "rapid manufacturing" to "additive manufacturing." When the more tabloid term "3D printing" was used is hard to determine, but it seems like the term at least was mentioned already in 1989 by Terry Wohler. In 2009 these technologies were labeled with an international standard (ISO/ASTM 52900) and revised in 2015 (among the terms described is "3D printing"):

2.3.1
3D printing
fabrication of objects through the deposition of a material using a print head, nozzle, or another printer technology
Note 1 to entry: Term often used in a non-technical context synonymously with additive manufacturing (2.1.2); until present times this term has in particular been associated with machines that are low end in price and/ or overall capability.

ISO/ASTM 52900:2015(en)

As seen in this paragraph, 3D printing is narrowed down to low-cost desktop printers. However, the industry, the Gartner group, and the media use the term "3D printing" to cover all technologies that produce 3D artifacts from a digital file. A wider term that is currently gaining traction is "digital fabrication."

That said, this book is not about the technology; it's about the uptake, use, and impact of the technology, and this spans more than two decades. Our interest is in design and its relation to these emergent technologies, not primarily as manufacturing but product design. As my colleague Andrew Morrison, my doctoral supervisor and an author of this collection, recently pointed out, we are actually in the business of additive designing in the context of digital fabrication. He argues that this includes interaction, systems, and service design, but our focus here is still on the designing of products. Consequently, I have chosen to use the different terms throughout the book, meaning you will meet rapid prototyping, rapid manufacturing, additive manufacturing, 3D printing, digital fabrication, fully freeform fabrication, and even other possible terms. Hopefully the contexts in which the terms are used in will provide better descriptions of the need and use rather than merely aligning with more technically correct ones not located in respect to design practice and analysis.

Since the arrival of our first machine, more than 1000 students have used a variety of additive manufacturing devices as part of their design education. I have had the privilege to work with almost all of them and to see their engagement and growth as this technology has also matured over time. My pedagogical approach has been one of dialogue and experimentation, querying and teasing out the possibilities of the technologies and processes of making and reflecting. Our work has always been curious about the possibilities these technologies offer design in the future.

Such a stance was also central to my own doctoral research completed in 2013. The design and pedagogical experimentation of working with additive manufacturing in its emergence was reflected in my own practice-based design research studies. Mine was one of the few early approaches to additive manufacturing that originated in design and, more specifically, product design. My thesis aimed to investigate how a product designer might approach and utilize additive manufacturing technology.

In this book I have drawn together the views and experiences of colleagues in design at the Oslo School of Architecture and Design and others. The chapters cover a range of design-centered views on additive manufacturing that are rarely addressed in the main conferences and publications that are still mostly, and importantly, concerned with tools, technologies, and technical development. Our interest is to try to locate and elaborate a design-centered view on additive manufacturing. The chapters are a mix of expertise and experience, some single-authored and several co-authored, reflecting dialogues about transdisciplinarity and the inclusion of domains such as business and aesthetics, narrative, and technology critique, and for us, two of the chapters are part of a second doctorate in additive manufacturing that is being conducted in the thesis by publication or compilation mode. I, too, have included my own more recent views and reflections, and I trust that together these will provide readers with a clearer design-centered view on additive manufacturing that might enrich a wider understanding of how design may inform technology development, use, and critique.

This book will hopefully interest other design educators and students, professionals, and researchers in design and technical schools, colleges, and universities and their engagement in related courses.

Steinar Killi
Oslo, Norway
March 2017

Chapter 1

Scope of the Book

Steinar Killi
Oslo School of Architecture and Design, Oslo, Norway
Steinar.Killi@aho.no

1.1 The Magic of 3D Printing

In 2003 we were contacted by Norway's national television broadcaster (NRK), which had a weekly science program called *Schrödinger's Cat*. NRK wanted to make a short feature of this new technology that could make 3D objects fast and accurately; if we could include 3D scanning at the same time it would be fantastic. We took upon us the challenge and suggested we could scan in the TV host and then make a physical model of him. We had a Minolta VIVID 900 3D Scanner and three selective laser sintering (SLS) machines to build the model in. We made a platform that could rotate 360° and placed the TV host on it; then we scanned him in one go, rotating the platform six times. The photographer was enthusiastic when seeing the red laser beam scanning the face and torso of the host (see Fig. 1.1 left panel). We bid the TV team farewell and promised a physical model of the host the next day.

The following night was probably one of the worst ever; the digital models from the scan were far from perfect and needed

Additive Manufacturing: Design, Methods, and Processes
Edited by Steinar Killi
Copyright © 2017 Pan Stanford Publishing Pte. Ltd.
ISBN 978-981-4774-16-1 (Hardcover), 978-1-315-19658-9 (eBook)
www.panstanford.com

extensive fixing. At the same time we had a clear deadline for when the machines had to be started. I will spare you the details, but the whole team made an oath never to try this kind of stunt ever again. Easy to see how these things are forgotten. Somehow, we were able to make a solid digital model and decided for the sake of a good TV effect to make the model in sand material, mixed with wax, which was the medium keeping the sand together. When the TV crew arrived in the morning, the machine had just finished and it was time for unveiling the result. We took a chance and opened the sand package with the camera running (see the still picture in Fig. 1.1, central panel). To our extreme delight and the host's childish amusement, his head and torso were slowly unveiled, sand drizzling down his cheeks and shoulders. "Like a troll coming out of a mountain!" was his enthusiastic comment (see Fig. 1.1, right panel).

Figure 1.1 Left panel: TV host Sverre Asmervik being 3D-scanned for a science program *Schrödinger's Cat*. Central panel: The 3D-printed head of Sverre Asmervik emerges from the powder bed were it's been printed. Right panel: The rather strange experience of holding a 3D-printed bust of yourself. Pictures: Screen dump from Norwegian Broadcasting.

Luckily few people know what the backs of their heads really look like; we had to perform some radical plastic surgery on the back of his head, adding hair where it had just been a hole, and so on. The shoot became a success, and the host probably still has his bust stacked away somewhere. Today this is a "walk in the park" and done at fairs and conferences in a fraction of the time. Actually, in London at the iMakr Store, you could go and do it yourself!

This little anecdote illustrates, however, just how far we have gotten in these 14 years, or is it that far? And, how could this, still intriguing, technology evolve further, and what will it become?

3D printing, or additive manufacturing (AM), has been predicted to initiate several changes of paradigm. Over the last 7–8 years, newspapers and TV shows have filled their editorial space with an increasing amount of technology-positivistic news about this alleged disruptive technology. Conferences dealing with 3D printing are popping up all over the world, and politicians are praising it to be the solution to almost anything—from bringing production back to a postindustrialized society to enhancing people's creativity.

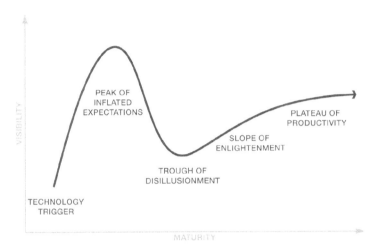

Figure 1.2 Gartner's generic "Hype Cycle for Emerging Technologies," visualizing the emerging technologies' journey from immaturity to an expected level of productivity. Picture: Courtesy of Gartner.

Gartner's "Hype Cycle for Emerging Technologies" is a yearly presentation of emerging technologies and their place in the evolution from what could be labeled a concept until it's matured enough to become productive (see the generic hype cycle in Fig. 1.2). The first hype cycle was introduced in 1995, but first in 2007 the concept of 3D printing was introduced. In 2015 the term "3D printing" was split into two: enterprise 3D printing, which has reached the "slope of enlightenment" and is expected to enter the last stage and the "plateau of production" within 2–5 years, and consumer 3D printing, which is still at the "peak of inflated expectations." Then in 2016, 3D printing surpassed the hype cycle.

Given this rather optimistic forecast, it's taken 30 years from when the first 3D printers were presented at an auto fair until it has reached a level of impact probably not anticipated by the inventors, who mainly developed a technology to rapidly produce prototypes and models of a design, hence the initial name "rapid prototyping machines." Actually it was not before the beginning of the millennia that the thought of using it for end user production emerged fully [1]. Probably these early machines were used to produce spare parts quite early, and the thought of using these for more than just prototypes evolved several years before.

However, there is a slightly more complex story behind the transformation from a model-making technology to the expectations we see today. Why did not Gartner pick up the technology before 2007, 20 years after Chuck Hull presented the first 3D printing device? What happened in 2007 that placed 3D printing within reach of the Gartner's hype cycle radar? Probably several things; I will discuss some of them since they point toward some of the challenges and possibilities 3D printing could encounter and unleash.

1.2 Legal Issues

I first came into the scene of rapid prototyping in 1994, when the Institute of Industrial Design at Oslo School of Architecture and Design (AHO) purchased its first prototype made with stereolithography (SLA) (see Chapter 2, about technologies) of a ski pole basket (see Fig. 1.3).

Figure 1.3 3D-printed ski pole basket from 1994. Design: Geir Eide; photo: Birger Sevaldson.

Then the school invested in a used SLS2000 from DTM (later merged with 3D Systems) in 1997. This was at a time when the Wohlers annual report (more about the Wohlers report in Section 1.6) listed all machines sold worldwide and how many machines the different countries had. Like with previous technologies, for instance, the video system, which came in three versions—Video Home System (VHS), Beta, and 2000—there were three technologies emerging at the same time, and there were also competing companies within the same technology. These different technologies will be broadly presented and discussed in Chapter 2. For instance, both DTM, an American company started by the inventor of SLS, Carl Deckert, and EOS GmbH in Germany had a similar technology—sintering a plastic powder with a laser. There were several legal issues between those two companies about patents on hardware and also on consumables (plastic powder). At some point during the early 2000s the same powder could be purchased in Europe at half the price as in the U.S., since a court ruling shut EOS out from the American market, but DTM (now 3D Systems) could do business in Europe. Along the fights between the big companies, several small enterprises challenged patents and launched competitive versions of the existing technology. In general most of these new companies were shut down due to the illegitimate use of patents. However, in 2009 many main patents, especially on the fused deposition modeling (FDM) technology, would expire and many small companies started to develop their versions of this technology, ready to be launched when legally possible [2]. So, the first affordable machines hit the market in 2009; in the following months and years several new companies entered the stage. A fascinating story about two of the most significant companies emerging in this period are depicted in a documentary called *Print the Legend*. Here we follow the birth and growth of Makerbot, from the start of an idealistic business idea with open source and open development as the founding pillars. Over time the open source idea was abandoned and Makerbot was eventually purchased by Stratasys, the founder and original patent holder from 1989. How many different companies are selling this technology today is probably hard to tell, but an estimate could be a couple of hundreds. The other company depicted in the documentary, Formnext, chose to develop a low-cost SLA 3D printer, also along the same paths as Makerbot. They met several obstacles around patent issues held by the inventor of SLA, 3D Systems. This company, however, is still an independent one.

Now, 10 years later, patents are expiring all the time. Some of the other original technologies have been challenged, too, for instance,

SLS by Laser Sinter Service, Prodways, and others. However, it was worth noticing that the low costs marked in 3D printers were totally dominated by FDM technology, a technology very close to what was described in the patent application almost 30 years ago! So, even if the patents have expired and low-cost 3D printers developed for hobbyists and the home market can be bought almost everywhere, it's still quite far away from being the alleged game changer.

There are other legal issues emerging along this technology now—copyrights and challenging of proprietary rights [3]. Both are similar to what we have seen in the music and film industries; existing products could more or less easily be 3D-scanned and reproduced, or brands could be copied or altered. This is again nothing new. Famous brands like Rolex and Lacoste have been copied and sold for almost as long as the brands have existed. Solutions are being offered, for instance, streaming of data to your printer, like Netflix.

What is quite clear is that there will be an increase in lawsuits within the 3D printing industry, among both the hardware (technology) providers and the users of 3D printers [2, 4]. This book will, however, not go into these issues.

1.3   The Power of Rhetoric

As mentioned before, the first 3D printers were labeled as rapid prototyping machines. When I started to attend rapid prototyping conferences in 1997, the technology and how to tinker with the technology were the main topics. The inventors and developers of both SLA and SLS in North America started user groups that held annual conferences; these were open only to people owning the technology but served as an early form of open source. As a user of the technology you would benefit tremendously from these conferences. When DTM (producer and inventor of SLS machines) merged with 3D Systems in 2001 the two user groups eventually merged, too. This user group went through some name changes before the final name "Additive Manufacturing User Group" (AMUG) was adopted; this was in 2008. Interestingly, this group, consisting of people who have been working within this industry since the early beginning, insists on labeling the technology as AM, a definitely more accurate name describing the process but definitely not very tabloid. At about the same time low-cost rapid prototyping equipment started hitting the market and the term "3D printing" hit the media. Also in 2007 the technology, the same as had been around for 20 years, suddenly emerged on Gartner's Hype Cycle for the first time but now with the name "3D printing."

From personal experience attending these conferences every year since 1997, I can say that the term "3D printing" does not fit well with how these "founding fathers" see the technology, again probably because the term "3D printing" promises more than what those low-cost FDM machines deliver. However, the term "3D printing" is so much easier to comprehend, explain, and imagine around that it is a losing battle.

1.4   Maturing of Technology

In 1928 the first mechanical TVs were sold. The flickering images, a poor replica of what could be seen in cinemas, did not predict a bright future. The time line of the invention of the television goes much further back, to 1878, when William Crookes confirmed the cathode rays by displaying them on a tube. The rest is, as one says, history. Over (a long) time the technology matured. More people bought TVs. The technology improved in steps, and the content coming out from the TV improved, too. About 25 years later, in November 1954, a similar event occurred; the first transistor radio was launched by Texas Instruments and Regency Division of Industrial Development Engineering Associates. This was a huge success. Although the TR-1 was of inferior quality to the stationary systems back home, the benefits of portable music were so huge that people bought them anyway. The next step in the story of the transistor radio was the birth of one of the biggest brands within consumer electronics— Sony. It launched its model TR-55 in the spring of 1955 and made a leap in quality. The company, originally named Tokyo Tsushin Kogyo, changed the name to Sony, a much easier name to pronounce in the U.S. The salesmen of transistor radios could be recognized by their shirts, which had extra-large pockets, allegedly big enough to hold a radio; these salesmen were called "sonny boys" [5].

This led to a business with a growing market for improvements. The key factor here is a disruptiveness that carries the product through its infancy and allows time to mature the product through an evolutionary process. It is, however, important to remember the sociocultural impact the 1950s had on the portable transistor radio. The 1950s were a decade where an emerging group of consumers with more spare time than those in previous decades lifted the portable transistor radio out of the private home and took them to the public room, the beach, etc. [6].

Very often quite simple and intuitive products have matured over a significantly long time. In "The Best Laid Plans of Mice and Men: The Computer Mouse in the History of Computing" [7] Paul Atkinson describes the long and complex road from the first concept of a selection tool for a computer screen, in 1963, that had some resemblance to what we today call a computer mouse as a product. Although this has been an undisputable commercial success and today the mouse is an indispensable tool, it took several steps and setbacks before this product reached its full marketable functionality and fame. One reason for this was that it involved an engineering process that lowered the production cost from US$350–400 to US$20 [7]. Also involved was the need to overcome perceptions and develop affinity to new ways of operating a computer, people being used to operating their actions and tasks from a keyboard. A massive effort was needed to convince and teach people about this device. Interestingly we now move more and more toward what is intuitive, with small portable devices controlled by the fingertips.

Further, TVs, transistor radios, and now the Internet and smartphones are all major technology breakthroughs that changed people's lives. However, learning to use them was easy. TVs and radios were/are close to plug-and-play, and the Internet and smartphones are close to being self-explanatory. The computer mouse was actually more difficult to adjust to. So, what about technology that actually requires quite a lot from the users to become similar game changers [8]?

1.5 "We Are All Designers"

This statement originally comes from Donald Norman [9] and has later been used by, among others, Chris Andersen in *Makers: The New Industrial Revolution* [10] and John Hockenberry in a TED talk of March 2012, who used it to predict a future where we all will be makers due to low-cost 3D printers, digital design tools, and a global sharing community, the Internet. This could probably be discussed. Norman argued that we create bonds to a product, thereby making it our own, but that does not really make you the designer of the product itself. Andersen correctly refers to the lower risk in starting up the production of physical products compared to 5–10 years ago. The web is filled with digital files of more or less useful stuff, available at production sites like Alibaba, which outsource production through low-cost manufacturing facilities, and Shapeways, which specializes in 3D printing parts, either from its own catalog or from files sent to it. In major cities, like London, there are also print shops located in the center of the city (see Fig. 1.4).

Figure 1.4 Window exhibition of an Imakr store in London, January 2016. Picture: Steinar Killi.

The question, however, is, does access to tools like 3D printers and computer-aided design (CAD) programs enable you to create, design, or make things with enough quality to actually be desired? Professional designers at, for instance, IDEO and Everett McKay at UX Design have questioned this statement:

"Welcome Everyone into the Design Process . . . Then Kick 'Em Out."
—Everett McKay

By stretching the term "everyone is a designer" to include opinions from stakeholders like executives, market analysts, focus groups, friends, and spouses, who could give structure and direction and help to choose among different concepts, the term is valid. However, to actually perform the design, in other words, giving shape and substance to a product, some specialized skills are necessary.

After 22 years of teaching on a master-level design school it has become obvious to me that giving form to anything, whether it is a 2D graphical expression or a 3D artifact, a high level of both skills and a willingness to work hard is mandatory! This does not mean that not everyone is creative, but luckily we excel in different arenas—some in making food, others in music, and still others in painting, gaming, gardening, writing, etc. Owning a saw, a hammer, and planks does not indicate you could even build a simple bookshelf, although many of us could be quite handy with guided training.

This ranting could maybe look a bit out of place, but this is actually the core of this book. There is anticipation from too many that just because the production of parts has been democratized and lowered the cost for one-offs and small batch production, it will lead to household designing and production of products, whenever needed. It is quite easy to see these as an obstacle to the household 3D printer, but it impacts the whole industry of making. When production shifts from a large series to a small series, it influences the process of developing products, including legal issues and the economy. Chris Andersen and others narrate a titillating story where small batch productions could be moved closer to the consumer and enhance the quality of the products, making them customizable and addressing small niches in the market. This is a story with a powerful potential and definitely within grasp, but what is bridging the market and this new way of producing is the *design* of the actual products.

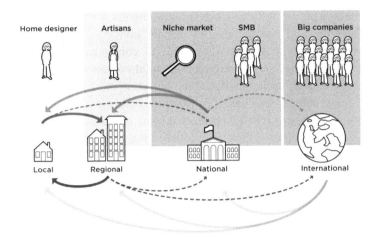

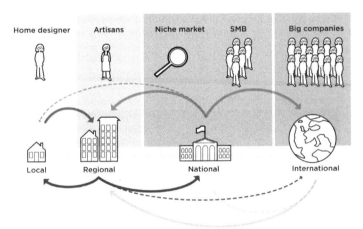

Figure 1.5 Possible scenarios of 3D printing along the line of producers. The design could be developed by all the stakeholders along the line; these designs could be picked up by all the stakeholders and be produced as well. However, it is not likely that all stakeholders will interact, as depicted in the drawing. The upper panel shows how the production could be distributed between the different stakeholders. The lower panel shows how design could be distributed between the different stakeholders. Drawing: Inger Steinnes and Steinar Killi.

If we imagine a line (Fig. 1.5) going from a low volume to a high volume from what could be called a prosumer, the home designer who creates, produces, and uses his or her own design to a big company's that churns out new products constantly, for instance, Adidas, IKEA, or BMW, we have a span of possible product development processes.

Somewhere between the two extremes on the line we have artisans, craftspeople, and companies that totally outsource the designs of new products to smaller niche market companies. Figure 1.5 visualizes how design and production could coexist and influence and create new opportunities. The big companies—automotive companies, companies dealing in wearables and electronics, etc.— have been using this technology since it came to the market in 1988. For these companies it was an excellent tool, in combination with other modeling techniques or just hand drawings. These companies continued using the technology when possibilities like rapid tooling emerged, also in the early 1990s. Most probably they experimented with a small series, testing out the potential for end user products, and they still do this. See Fig. 1.6 of a 3D-printed sneaker from Adidas, FUTURECRAFT 3D; just a few couples of this sneaker were made. However, in spring 2017, Adidas launched the second generation, FUTURECRAFT 4D, in collaboration with the 3D printing company CARBON, a different technology than that showed in Fig. 1.6. This version was planned to be produced in a much higher scale, by the thousands!

Figure 1.6 Producing the limited series of sneakers for Adidas, FUTURECRAFT, with a 3D-printed sole. The outsole is made of textile with a more standardized method. Picture: Adidas. FUTURECRAFT and the 3-Stripes mark are registered trademarks of the adidas Group, used with permission.

In 1998 we started out a research project with four semisized Norwegian companies: Hamax, a producer of snow sledges in plastic, which bought its design from external design offices; Luxo, a producer of lamps, mainly for desktop use and with in-house design expertise; Polymoon, a producer of mainly packaging for food, with both external and internal design resources; and Iplast, an injection molding company that went bankrupt during the project and was replaced with Jordan, producing dental products, mostly toothbrushes. The project sought to see how rapid prototyping would influence the design process, with the hypothesis that physical products during several stages through a product development process would act as catalysts in multidisciplinary meetings, reducing the risk of developing new products, and thus lowering the risk of lost investments in production equipment, which could run up to millions of dollars. The project ended in a PhD where different types of prototypes for the different phases of product development were presented [11]. This will be further elaborated in Chapters 3 and 4. During this project not only prototypes were made but also jigs for stacking during production and simple tools for test production were made, but the intention was never to look into the actual production by 3D printing; more about this in Chapter 4.

So, the AM technology has been used by big companies and smaller ones, including design offices, for decades already. They are familiar with the technology and are of course on the lookout for a technology and an application that would make it worthwhile to 3D-print at an industrial scale. On the path to this, there will be development in how such products will be developed—with this I mean both the technology and whatever the application will be there will be other requirements to how it is made. Today there exists a large amount of literature, YouTube videos, and courses that show you how to design for 3D printing. This is only half true: They show how the technology influences the design—the different wall thicknesses possible, the complex overhangs possible, etc. (more about that in Chapters 2 and 3)—but they rarely talk about how the application (product) itself could be developed when producing for 3D printing. So, what goes for the household 3D printer designer also has relevance for all parties along the line, from households to big companies (Fig. 1.5). Why this is of special importance will hopefully become clearer in the next section.

1.6   From Rapid Prototyping to 3D Printing

There are several excellent books, articles, webpages, etc., describing the invention and development of 3D printers. The rather short version is about a technology that came out from two other emerging technologies, CAD and robotics. The different technologies we see today are largely the same as those developed toward the end of the 1980s. They are bigger and faster and of higher quality, but in general, it's the same principles that still rule the business. The different technologies are presented in Chapter 2, so if the technology is unfamiliar, I would suggest reading through that chapter before continuing.

It is quite obvious that similar to, for instance, the development of the computer mouse, 3D printing has evolved and matured over several years, but also similar to other, previous emerging technologies, there are several reasons why a technology converges from a niche market to something global. In the case of 3D printing one of the main converging reasons was patents going obsolete. In 2009, about 18 years after patents for FDM (see Chapter 2), low-priced printers utilizing this technology hit the market. How many versions/companies exist on the market today is probably impossible to state, but there are hundreds. In 1995 the first issue of Wohler's report came out; it has become an annual publication that gives an overview of what has evolved from the rapid prototyping industry to today's AM and 3D printing. In the first years, from 1995 until 2001 this report provided an oversight on not only how many 3D printers had sold during a year but also how many each country possessed. For instance, in 1998 there were four registries of rapid prototyping machines in Norway [12]. In a user group conference in 2002 one of the discussions was if the market was about to be saturated with rapid prototyping machines. It is also quite obvious that the reason for the explosion in new companies developing 3D printers and the following hype was not the possibility to make models and prototypes; it was the possibility to make useful products fast and relatively cheap.

There are, however, some obstacles on the road to the alleged third or fourth Industrial Revolution. Some are of a technological kind, but there are other, subtler obstacles that have to be addressed on this path: There are legal issues, like in the music and film industries, the digital format raises some obvious challenges toward copyrights and piracy. There is the social context—are we ready/interested in taking the next step into a world that will be both defragmented and more global at the same time? 3D printing has often been discussed as the catalyst for the long-tail production paradigm [13]. Also, like all digital products and their outcomes, the quality of what goes in strongly reflects in what comes out, or as the old programming proverb goes, "Crap in, crap out." It is the latter part this book will address. The technology is obviously of great importance and will be implemented; however, there are better books delving into the specifics of the technology.

When going to what then was called rapid prototyping conferences in the beginning of the 2000s, one of the most used statements was, "We need to train the designers for this new technology!" Books like *Rapid Manufacturing, an Industrial Revolution for the Digital Age* [14] and Wohler's report have paved the technological road for 3D printing, but what they generally are saying is, "We have a fantastic technology; it should be used for something great." In many ways they are pointing to the *possibilities* that lie in this technology. Krippendorff [15] defined three directions in a product development process: *challenges, opportunities*, and *possibilities*.

Challenges are what we earlier would call problem-oriented views. For many designers it is important to define a problem they can solve. Obviously, 3D printing is a strong problem solver, whether it is to produce spare parts or to produce any highly complex part. *Opportunities* are what we could call solution driven/oriented. This approach is thought to be more holistic than the previous one. An example here would be the hearing aid industry developed through 3D printing. *Possibilities* are not necessarily connected to a specific situation where there is some kind of need that should be addressed. This includes design offices like Freedom of Creation that become inspired and then design a product, not necessarily for a company or even a defined customer. In Chapter 3 there is an example of possibility-driven design, a structure made up of aluminum tubes and 3D-printed joints. It was designed to show the possibilities with 3D printing in, for instance, architecture. Of course, very few designers would define themselves completely within one of the three groups described; however, we could use this demarcation to systemize different approaches to product development.

Now, if we follow the technological track, it starts and ends with the actual production; you start with the digital representation of a physical product that could be given a physical form through a toolless process, adding layers by layers, hence the name "additive manufacturing." The University of Loughborough made in 2004, later also used by Chris Andersen [9], a graph showing the cost-per-part comparison between different 3D printing technologies and injection molding (Fig. 1.7). The graph underlines what has become the repeating mantra; since there is no investment, the price per part does not go down if you produce a larger volume; in other words, it could be beneficial to produce unique products, whether they are customized or just extremely specialized and require a very low volume.

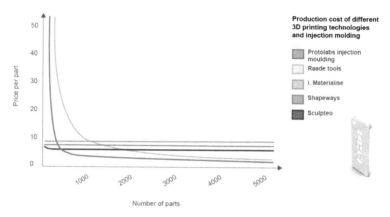

Figure 1.7 Comparison of production cost, for a cell phone cover, between different 3D printing technologies and injection molding. Picture: Henriette Marki and William Kempton.

This model is unfortunately somewhat misleading. It adds in terms of tooling cost for injection molding, leading to an extremely high cost for a low volume and then a rapid decline when the volume grows. However, when developing a new product, investments do not just cover tools; the development cost will also be distributed to every part made. In other words, if the part produced is the only one we need/want and the production cost is 5 euros, the development cost would probably exceed that amount by several times. We took upon ourselves to investigate a more realistic curve with the development cost and investment cost added in. A quite simple phone cover was designed (same as that shown in Fig. 1.7), and the tooling/production cost was obtained from a tooling/production company. Further, the file of the cover was sent to Shapeways for a quotation. A new graph with the different cost was plotted (see Fig. 1.8).

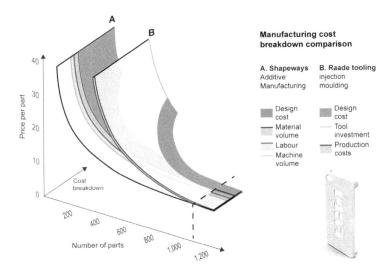

Figure 1.8 Breakdown of actual cost, including development. Comparison between 3D printing and injection molding. Picture: William Kempton and Steinar Killi.

What it shows is the significance of the development cost, and furthermore, what happens before the actual production (printing or molding) is probably even more relevant for 3D printing than for standard manufacturing methods like injection molding. There is a "sweet spot" in the curve that lies between 200 samples (below that number, each cover would be indecently expensive and in no way competitive even though it is custom made for a university) and 1050, where it would be more cost beneficial to produce it with injection molding.

This tells us that what happens before the file goes for printing; the actual design work is of utmost importance for this technology to achieve its potential.

This book is meant to be a contribution to the process of bringing 3D printing further beyond the last stage of Gartner's hype cycle, the plateau of production, addressing the gap and building a bridge between the new way of producing and the anticipation of a new product development paradigm.

The different chapters will deal with the different stages in new product development, giving both a holistic input to the design process and some more hands-on methods suited for this technology.

1.7 The Chapters of the Book

Since production technology—3D printing or the more correct name, additive manufacturing—is one of the main reasons for this book, Chapter 2 gives an account of the social and technical phenomena surrounding the evolution of 3D printing, seen from the perspective of design and product development. There is also an appendix presenting different additive manufacturing technologies. Chapter 3 is the central chapter of this book; a model for product development utilizing 3D printing through the whole process, from conceptualization to production, is presented and exemplified. This will lead to Chapter 4, reflecting on the impact the technology has on the industry today, from prototypes to production. Chapter 5 then delves into different methods for generating form. Chapter 6 presents models of branding and why this is of increasing importance in all product development today; and for products designed for small niches or to be customized, it is, if possible, of even more importance. Finally, this book aims to discuss the future impact of 3D printing; the final chapter seeks to challenge some of the paradigms presented both in this book and others.

References

1. Wohler, T., ed. (2004). Wohlers report 2004. In *Wohlers Report*. Fort Collins, p. 270.
2. Hornick, J., and Roland, D. (2013). Many 3D printing patents are expiring soon: here's a round up & overview of them (accessed on May 11, 2016).
3. Lucas, M. (2015). The future of 3D printing: smarter IP strategies, less lawsuits (accessed on May 11, 2016).
4. Earls, A., and Baya, V. (2014). Technology forecast: the future of 3-D printing, p. 2 (accessed on May 11, 2016).
5. Vintage Transistor Radios (2016). Secondary vintage transistor radios [Web page].
6. Sandström, C. G. (2010). A revised perspective on disruptive innovation: exploring value, networks and business models (Doctoral thesis). Chalmers University of Technology.

7. Atkinson, P. (2007). The best laid plans of mice and men: the computer mouse in the history of computing. *Design Issues*, **23**(3), 16.

8. UTK, T., ed. (2015). The 3D printing game changer.

9. Norman, D. A. (2004). *Emotional Design: Why We Love (or Hate) Everyday Things*. New York: Basic Books.

10. Anderson, C. (2012). *Makers, The New Industrial Revolution*. New York: Crown Publishing.

11. Capjon, J. (2004). Trial and error based innovation (PhD thesis). Oslo School of Architecture and Design.

12. Wohler, T., ed. (1999). Rapid prototyping and tooling, state of the industry. In *Association*. Fort Collins, p. 265.

13. Anderson, C. (2006). *The Long Tail: Why the Future of Business Is Selling Less of More*, 1st ed. New York: Hyperion Books.

14. Hopkinson, N., Hague, R. J. M., and Dickens, P. M. (2005). *Rapid Manufacturing: An Industrial Revolution for the Digital Age*. Chichester: John Wiley and Sons.

15. Krippendorff, K. (2007). Design research, an oxymoron? In Michel, R., ed., *Design Research Now-Essays and Selected Projects*. Basel: Birkhauser Verlag, pp. 67–80.

Chapter 2

A Design Sociotechnical Making of 3D Printing

William Lavatelli Kempton
Oslo School of Architecture and Design, Oslo, Norway
william.kempton@aho.no

2.1   Introduction

2.1.1   Disciplinary Boundaries and Claims to 3D Printing

As 3D printing is moving toward the attention of wide-ranging societal context and spaces, it is taking on an increasing amount of roles and potentialities [25]. While initially seen as a developmental tool for designers, architects, and engineers, digital fabrication (with 3D printing as a lead indicator) is moving outside of these professional practices. Its recent claims span from a facilitator of distributed and personalized material production [5] to future challenges for intellectual property [7]. It is seen as a tool for self-motivated makers [1] to turn digital information into physical reality and as a platform for free, open-source innovation [42]. While the

Additive Manufacturing: Design, Methods, and Processes
Edited by Steinar Killi
Copyright © 2017 Pan Stanford Publishing Pte. Ltd.
ISBN 978-981-4774-16-1 (Hardcover), 978-1-315-19658-9 (eBook)
www.panstanford.com

development of digital fabrication and 3D printing has been ongoing since the 1980s [44], the popular and academic interest in this emerging technology can be demarcated by the expiry of major 3D printing patents in 2009 [6].

In engineering-driven disciplines [13, 15, 17], much emphasis is put on a particular avenue known as additive manufacturing (AM), which can be seen as advancing the concept of rapid prototyping (RP) in a mass-manufacturing paradigm. RP has since the early days of 3D printing facilitated the rapid making of prototypes for product designers, engineers, and architects. Here, 3D printing in the context of design is situated close to developmental methodologies, typically as a means of confirming ergonomic, visual, or mechanical considerations within a design space [33].

Whereas RP is often linked to developmental methodologies, the logic of additive material production moves us closer to the nature of the fabrication process. As opposed to subtractive material production, such as vacuum forming, injection molding, or die casting, additive fabrication implies the accumulation of matter into layers and building blocks, thereby evoking an idea of fabrication inspired by nature [31]. In relation to the potentials of AM in a consumer-oriented version of design, it gives the ability to make complex, customized, and multimaterial artifacts, while still being "cost effective" and "giving the potential for much greater customer satisfaction" [17].

As opposed to mere consumption-oriented digital fabrication, Gershenfeld (2008) relates these technologies closer to personal, developmental use [14]. Gershenfeld remarks on how having access to 3D printers, laser cutters, and other manufacturing tools creates a "physical notion of literacy." In this sense, Gershenfeld likens literacy to the ability to express oneself through whatever means are available.

2.1.2 Introducing a Sociotechnical Perspective to 3D Printing

At its core, this book chapter analyzes the emerging social practice, mediation, and knowledge that are being carried through the development of the 3D printer. This argument is based on a social constructivist idea of technological development as being constructed through social intervention. I align my arguments to the critical theories of science and technology studies (STS), such as those by Feenberg, Ihde, and Bijker, in order to contextualize man-made technological artifacts into their natural, technical, and social environments. This argument is facilitated through a set of theoretical concepts, such as the relevant social groups that interact with and mediate the use of technological artifacts through a technological frame. This argument of sociotechnical interplay allows my discussion to center on the production of knowledge, as opposed to the decontextualized technological production of artifacts.

By incorporating theories and discussions from sociotechnical studies [4, 11, 18], I take on an ontological view of technological development as being undeterministic. Such a view implies that nonintentional use and technological adoption in the "real world" influence the way technology is constructed. This view can be seen in contrast to linear, deterministic views of technological development that imply that it is decontextualized from its users and placement in societal contexts.

2.1.2.1 Sociotechnical development from a design perspective

Through the disciplinary views and uses of digital fabrication that I will discuss throughout this chapter, claims are being made that reposition boundaries between digital information and physical material making. The unpacking of disciplinary perspectives of digital fabrication is also relevant to both design research and practice, such as in human–computer interaction (HCI), whose concern with digital fabrication involves developing tools that support making processes [24]. As a technology that facilitates the making of physical artifacts, digital fabrication's influence on product-oriented design disciplines is wide ranging. When considering product design as an iterative "problem solving" process that involves stages of trial and error [38], digital fabrication allows designers to prototype and imitate abstract design concepts into concrete material visions, both quickly and efficiently.

Although digital fabrication is mostly associated with the production of plastic, metallic, or ceramic goods for conceptualizing and making artifacts, it is also possible to fabricate with novel materials, such as edible food. The 3D-printed gingerbread house, shown in Fig. 2.1, seeks to illustrate one of many potential engagements with 3D printing, which brings into question a view of digital fabrication, specifically that of 3D printing, as a platform for making "useful" artifacts. The digitally fabricated gingerbread house problematizes new avenues for engagement with novel materials and digital fabrication that are found on the periphery of conventional, consumer-oriented design.

Figure 2.1   Novel material approaches to 3D printing through the making of a digitally fabricated gingerbread house. Photo and design: William Kempton.

Through contexts and sociotechnological perspectives that are built up in a way that is relevant for design, I ask, "What are the perspectives and disciplinary claims that allow us to understand a contemporary view of 3D printing?"

2.1.3   Outline

The chapter is separated into seven sections. The first section introduces views of design and multidisciplinary views on digital fabrication. In the second section I elaborate on the concept of sociotechnical development as an undetermined process, to be played out by several relevant social groups. This is then contextualized within my discussions of digital fabrication and theories of nonlinear technological development. The argument for a socially constructed perspective of digital fabrication and making is then furthered through the introduction of several disciplinary perspectives. In Sections 2.3, 2.4, 2.5, and 2.6, I describe the development of 3D printing from the respective perspectives of 3D printer inventors, business managers, designers, and layperson makers. In Section 2.7 I discuss and summarize my conclusions.

2.2 Socially Constructed Technologies and 3D Printing

2.2.1 The Relevance of Social Groups

The engagement of people has an influence on the shaping of technological artifacts. This is particularly evident in the case of the development of technologies for home video entertainment. Since the introduction of the first video cassette recorder (VCR), followed through with the analog Video Home System (VHS) and Betamax format, and eventually Blu-ray versus HD-DVD, a series of format wars have occurred, sparked by social intervention. The Hollywood studios, the technology manufacturers, the government, and the users themselves can be seen as relevant social groups that all play a role in the shaping of domestic video technology.

In his monograph on the development of VCRs, Greenberg (2010) also emphasizes the importance of looking in between the "traditional protagonists" of sociotechnical development [16]. In the case of VCRs, Greenberg points to the way new technologies are mediated and the emerging contexts that surround it, what he describes as the "layers of mediation [which] help to package, distribute, and sell the product" [16]. In the early days of the VCR, small business owners mediated new technologies to their customers by renting out video cassettes and players. Although VCRs were initially intended for time-shifting (recording TV series while away), they also made it easier for the relevant social group of VCR owners to make their own movies and record footage using VHS-compatible video recorders. Early adopters of these technologies could record, transfer, and edit family videos and amateur footage at home, without having to go through the elaborate process of developing 8mm and 16mm film.

2.2.2   From Video Production to Material Production

The sociotechnical development of video entertainment changed many of the existing patterns of video consumption. Before the development of VCRs, owners of TVs were usually confined to the watching of direct TV broadcast at a specific time allotted by the broadcasters. Popular TV shows would be sent during prime time, accompanied by advertisements, which generated revenue for the broadcasters. Through new innovations in video technology, the VCR allowed users to materialize their own content by time-shifting their preferred TV shows, possibly skipping adverts, as well as video-recording and editing their own movies. These can be seen as acts of materializing content. Much in the same way, the concept of 3D printing involves actors materializing physical content by additive means.

2.2.3   Technologies for Additive Making

Figure 2.2 shows two design students in the act of producing material artifacts in clay and plastic using desktop 3D printers. While their attention seems to be focused on the desktop fabricators' remarkable ability to reproduce their design, the proximity of the computer in the background hints at the important presence of digital applications for facilitating the "making" of their design. The landscape of 3D printing, from large industrial processes to small, desktop fabricators is in fact made up of layers of technological systems. Engineered mechanisms such as high-precision stepper motors provide precise motion, while strings of computational code connect and put these mechanisms into useful motion. In a decontexualized view on technology, the sum of all these elements can be viewed as affordances that can be optimized and evaluated in terms of efficiency. In such a quantitative portrayal of technology, the success or failure of the various 3D printing technologies would be a simple matter of evaluating the technology with the highest output, quality, and affordability.

However, from a situated, socially aware point of view, the technological development of tools can be viewed as an undetermined entity, as it is always subject to use and manipulation by nature and society. As with any technology that exists or has ever existed, from hammers and nails to online sharing platforms, social interaction shapes the uses, contexts, and discourses on technology. This consequently leads technology onto unlinear tracks of development. In the case of digital fabrication and 3D printing, designers, engineers, makers, educators, and business developers contribute to different discourses on the same basic devices. While the current desktop 3D printers, currently capable of producing small-scale plastic artifacts, may be critiqued by professional actors within quality assurance and operations management as being of inferior quality to other industrial processes, their availability to an audience outside of organized product development is opening up paths for new entrepreneurial endeavors.

Figure 2.2 Design students involved in materializing clay artifacts with a desktop 3D printer. The computer in the background hints at the complex assemblage of technologies that 3D printing relies on. Photo: William Kempton.

2.2.4 Critical Theories and Studies of Technology

In his critical theory of technology, Feenberg and Callon (2010) analyze the construction of technology on two levels [11]. Firstly, technology is decontextualized from its users, situations, and things into basic technological affordances. Secondly, it is recontextualized into natural, technical, and social environments, which is essential in order to understand the real-life world of technology. Feenberg and Callon introduce the concept of technical code to discuss the rule under which technologies are developed in social contexts, with biases reflecting the unequal distribution of social power [14]. As technologies evolve, new social groups challenge the technical code through new designs. Such a case is famously described by Bijker (1997) in relation to the development of the bicycle at the turn of the nineteenth century, which fragmented into opposing bicycle designs used by different relevant social groups [4]. While the larger-wheeled bicycles, such as velocipedes, appealed to a completive social group, the smaller-wheeled safety bike was perceived as more utilitarian. Only after the invention of the rubber tire did the opposing social groups converge on a similar technological track by using smaller wheels. The story of the development of the bicycle serves as a well-suited example in explaining undeterminism as a hermeneutical track. One can interpret the inevitable stabilization of the now archetypal bicycle design as inevitable, due to functional improvements in the making of rubber tires. However, as is central to Feenberg and Callon's perspective, the concept of function has no use for meaning, as "the concept of 'function' strips technology bare of social contexts, focusing engineers and managers on just what they need to know to do their job" [11].

2.2.5 Unpacking the Views of 3D Printing

When the first additive digital fabrication technologies were developed in the 1980s, they were primarily used by engineers and designers for verifying ideas and prototypes in a product development process. The use of the term "rapid prototyping" gives an impression of the limited scope of use at the time. However, as the maturation of cheap, powerful electronics converged with the liberation of certain 3D printing patents [6], the landscape of 3D printing rapidly opened up to new fields of users and social contexts.

From a social sciences perspective, Birtchnell and Urry (2016) discuss 3D printing from the perspective of increasing globalization and mobility and how it might really reconfigure the existing patterns of production, distribution, and consumption. The authors point to great social-technical potentials of 3D printing, such as its ability to tailor for individual needs and use, which in turn might alter the current consumptive paradigms we live by [5]. Arguably these emerging technologies have the potential to change, or at least challenge, the current paradigms of consumption. However, I place the conceptualization of these changes in the hands of emerging design practice.

Broken down to its bare technological affordances, 3D printing has arguably not changed much for design practitioners since its original inception. Technologies that deliver high-quality artifacts, such as selective laser sintering (SLS) (see Appendix) and stereolithography (see Appendix), were in fact the principal technologies to be developed. What has changed significantly is the way the technology is approached by its relevant social groups. In the early years of 3D printing, the complex equipment and necessary expert knowledge demanded dedicated third-party service providers and large corporations. Designers and engineers would send 3D blueprints to third parties and have them sent back as tangible artifacts, allowing them to verify technical or aesthetical concepts at critical moments.

The process of making with 3D printing is today increasingly weaved into the initially conceptual, value-making stages of design. This has also made it simpler for smaller, less capital-heavy organizations, such as local maker spaces and design collaboratives, to partake in and use digital fabrication. New ideas, concepts, and services related to digital fabrication emerge, from novel household products to interactive robots that stimulate the presence for sick schoolchildren [29] and distributed manufacturing networks such as 3Dhubs.com.

2.2.6 Socially Constructed Perspectives of Additive Making

Following up on my initial argument for constructing a technological perspective that is aware of social practices, I will now turn to a contextualization of what 3D printing means from the perspectives of a set of relevant social groups. These have been identified as inventors, businesses, designers, and laypersons who are involved with 3D printing. While all of these social groups discuss and envision possible futures and potentialities of 3D printing and digital fabrication, their claims of knowledge are often conflicting. This can be seen in their perception of value creation, the way they discuss 3D printing futures, and their perceived ambitions.

Focusing on the actors and groups that interact with a certain technology enables a shift from the discussion of technological artifacts to a discussion of knowledge. In the case of the emerging use of VCRs in the 1970s, the relevant social groups were perceived as the technology makers—the television studios and the users of VCRs themselves. The honing in on the interaction between these social groups reveals how the VCRs not only facilitated predetermined uses, such as time-shifting (the recording of a show and watching it later) but made it easier for users to record and create their own movies. New technological mediators emerged, such as local video rental stores, which significantly impacted the way people could watch movies at home. Instead of going to the cinema, VCR owners could rent and watch movies in their own living rooms. The question then lies in what it means to be a part of a relevant social group. Also, how can we structure and present the opposing views and perspectives of the relevant social groups who interact with 3D printing as a technological artifact?

2.2.7 Relevant Social Groups as Part of a Technological Frame

To structure the interactions among actors in a relevant social group, Bijker (1997) introduces the concept of technological frame [4]. The technological frame is emphasized as a theoretical concept that is established around the interaction with a particular technological artifact. Elements such as goals, key problems, user practice, problem-solving strategies, perceived substitution function, and tacit knowledge are some of the elements that make up the technological frame. These elements might also vary, depending on which social group the technological frame belongs to. As to the question of who are the relevant social groups, Bijker emphasizes the importance of seeing all relevant social groups as being equally relevant, thus making it reasonable to include nontechnologists, such as layperson users, in my sociotechnical analysis of 3D printers.

The technological frame, as illustrated in Fig. 2.3, reveals how the various relevant social groups interact with each other and the technological artifact that is the 3D printer. The 3D printer inventors are concerned with improving and making new fabrication processes to improve the volume of production output, as well as to deliver consistent results. The social group of business managers strives for creating new value systems within a paradigm of economies-of-one, with key problems seen as the necessary operational restructuring of production and delivery. The perspective of the social group of designers is more entangled in exploring material, aesthetical, and process possibilities and in using 3D printing as a developmental design tool. The social groups of laypersons, often referred to as makers, is concerned with conceptualizing and making for private purposes. However, the open sourcing of 3D printing techniques is increasingly blurring the lines between the 3D printer inventors and laypersons. Laypersons are challenged with acquiring adequate proficiencies for material making, as well as having accessibility to 3D printing.

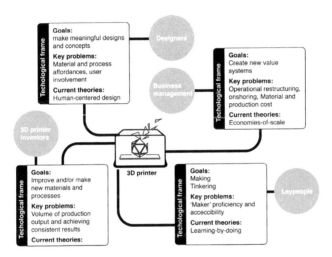

Figure 2.3 The relevant social groups relate to a technological artifact through a technological frame. Illustration: William Kempton.

2.3 The 3D Printer Inventors

The relevant social group of 3D printer inventors apparently forms a critical part of the development of the 3D printer—after all they consist of the researchers and engineers who develop and produce the tools used by those who acquire and use them. So why the need to discuss relevant social groups other than just the machine inventors themselves? The 3D printer makers continuously make new tools and improve the technical workings of the processes. But for whom are they making the machines?

In Chapter 1 Killi briefly discusses the relevant social groups of 3D printer inventors from a legal perspective, such as the German manufacturer EOS, which was banned from the American market due to licensing issues with DTM Systems over the use of printing materials and techniques. Also, as mentioned earlier, the expiry of Crump's patent in 2009 for fused deposition modeling (FDM) printing caused a wave of open-source, desktop-size 3D printers that were increasingly aimed at a consumer market. The developmental story of 3D printers is in fact riddled with legal issues—from the corporate battles of the 1990s and 2000s to the second wave of consumer-oriented desktop 3D printers from 2009 and onward.

2.3.1   The First Wave of 3D Printer Inventors

The initial invention of 3D printing processes, machines capable of automating the process of producing 3D objects by additive means, can be traced back to the 1980s. Spawning out of individual and university research projects, companies such as DTM Systems, Stratasys, and 3D Systems successfully managed to realize the early visions of physically reproducing digital blueprints. At the time, computer workstations capable of processing complex 3D graphics made it possible for engineers and designers to go from physical to digital drawing boards. Computer-aided design (CAD) programs such as Sketchpad had been around for some time and paved much of the way for interacting with digital 2D drawings.

With the development of a second generation of CAD programs, such as CATIA (initially developed by Dassault Systèmes for designing fighter jets), drawings could now be viewed and made in all three dimensions, as opposed to drawing-board-like 2D drawings. This development made it considerably easier for designers and engineers to design complex surfaces and geometries, which could then be visualized through computer rendering and simulations. However, the transition from visual to tangible artifacts still relied on laborious handicraft.

In an interview Hull [40] points to the costly and time-consuming process of developing prototypes and molds for plastic injection as a motivation for creating the first 3D printer. As he explains, the process of designing injection-molded plastic components would be a tedious process in the pre-3D printer era. A tool maker would craft a pattern from a set of technical drawings, which would then be cast into a mold. This process would often have to be redone, as either the original pattern or the molds wouldn't look like or separate as planned. As a result, the development time of even simple plastic objects could take months. Being an engineer himself, Hull envisioned a device that would automate much of the laborious work, which eventually turned into the concept of stereolithography—curing layer upon layer of photopolymeric resin using a scanner-aided laser.

The first generation of 3D printer manufacturers, such as DTM Systems, Stratasys, and Hull's 3D Systems, would continue to grow in the 1990s and serve an ever-expanding industry with specialist equipment for rapid prototyping. As the technology was costly, only large corporate structures could afford such investments, resulting in use mainly by the auto and aero industries. Smaller businesses, on the other hand, would have to resort to acquiring their services from model-making service bureaus. Statistical reports from Wohlers Associates [45] note that by 2004 most manufacturing industries had to some extent embraced RP. In 2003, the total sales of 3D printers amounted to 1864, which gives some indication that the availability of 3D printers was little and far apart. With "low cost" 3D printers such as the Stratasys Dimension SST selling for $25,000, this meant that RP equipment would be prohibitively expensive for smaller organizations, not the least for individual use.

The concept of RP soon came to be closely associated with all things 3D printing. Functional prototypes and aesthetical models accounted for approximately 50% of all applications in the mid-2000s. Although the glory cases were few and far apart, success stories such as the Siemens-developed hearing aids and Invisialign dental braces stand as rapid manufacturing success stories. In fact, Lipson points to the fact that 3D printers as early as the 1980s were sold as the "future of manufacturing" [25]. Killi similarly points out in Chapter 1 that the perceived application areas of these technologies transitioned quite fluidly between being prototyping tools and manufacturing applications.

2.3.2 The Second Wave of 3D Printer Inventors

With the expiry of the desktop-friendly FDM patent by Scott Crump and Stratasys (1992) in 2009 came a second wave of hype around 3D printing [6]. This time, however, the technology was not sold as a prototyping machine for corporate industry. Aside from the RepRap project, one of the first companies to create truly desktop 3D printers on a large scale, MakerBot announced its Thing-O-Matic 3D printer as a "cutting-edge personal manufacturing" tool [27].

Their creators, Bre Pettis, Adam Mayer, and Zach "Hoeken" Smith (Fig. 2.4), having developed the MakerBot concept out of the hackerspace NYC Resistor in Brooklyn, stayed in close association with the Maker community (The name of the company, MakerBot, gives some indication). Made out of laser-cut birchwood and held together with ordinary nuts and bolts, the Thing-O-Matic was itself advertised as an open-source "personal manufacturing" tool and sold as a user-assembled DIY kit for $1099. Boasting a build volume of 96 × 108 × 115 mm^3, the MakerBot printer relied on a plastic filament (either acrylonitrile-butadiene styrene [ABS] or polylactic acid [PLA]) to be heated up, melted, and applied to a build plate through a metal nozzle. 3D models, either self-made or downloaded via their own online repository Thingiverse.com, could be transferred to the printer via a SD card or USB connection.

Figure 2.4 Adam Mayer, Zach "Hoeken" Smith, and Bre Pettis in front of Cupcake CNC prototypes, the first MakerBot product. Image: MakerBot.

Although the company was later sold to Stratasys, which actually held the original FDM patents, MakerBot stood to symbolize the new generation of makers and 3D printer developers. Similar stories can also be told of Ultimaker, its European equivalent. Developed by Erik de Bruijn, Martin Elserman, and Siert Wijnia, who met at a RepRap convention in a maker space called ProtoSpace in Utrecht in 2011, the Ultimaker original shared a lot of the same traits. Constructed as a wooden, laser-cut chassis, the original Ultimaker was sold as a DIY kit for enthusiasts. Having a slightly larger build volume of 200 × 200 × 200 mm^3, the Ultimaker not only had a larger build volume but also a faster print speed due to its Bowden-driven extruder.

The second wave of 3D printer inventors can be seen as increasingly overlapping with the relevant social group of layperson users of 3D printing. Having met at hackerspaces/maker spaces, both the developers of MakerBot and Ultimaker started out by developing and manufacturing their products there, before moving to larger offices. Both companies also emphasize their community-based research structure, by open-sourcing both their hardware and software in online repositories. As a lot of the subcomponents of the fabrication devices consisted of off-the-shelf hardware and electronics (a lot of the desktop printers use Arduino prototyping boards), they were within the price range of curious tinkerers. While the self-assembled MakerBot Thing-O-Matic cost $1099, the Ultimaker was priced at €1194. Countless similar stories can be told of 3D inventor start-ups that have emerged since 2009.

How do the 3D printer inventors shape our understanding of the 3D printer in relation to design?

The first wave of 3D printer inventors positioned the 3D printer as a highly engineering-oriented tool, as it facilitated the manufacture of other things, through prototypes and as visual design representations (VDRs), giving it the name "rapid prototyping." As mentioned, there also existed early visions of making consumer artifacts directly with the 3D printer. However, it wasn't until the release of the desktop-oriented 3D printer by the second wave of 3D printer inventors that a similar vision became somewhat realized, although it was targeted toward tinkerers and hobbyist makers.

In the following section, focusing on the business perspective of 3D printing, I will further unpack how 3D printing is interpreted as a tool and means of innovative services and offerings for the pursuit of value creation.

2.4 Business Perspective of 3D Printing

From a business perspective 3D printing is often accompanied by predictions, critiques, and economic forecasts of how it could change future modes of manufacturing, distribution, and work life, ranging from highly positive to negative. Popular news media such as the *Economist* note that the convergence of digital software, new materials, and dexterous, distributed robots will enable a new generation of entrepreneurs to "start with little besides a laptop and a hunger to invent" [28]. Others present more balanced views where AM technologies will continue to improve and supplement, rather than replace current manufacturing paradigms. Sasson and Johnson (2016) envision scenarios where "manufacturers with complex bills-of-material will adopt 3D printing to extract additional scale advantages from traditional manufacturing" [39]. Some critical perspectives emphasize new copyright nightmares [25], while others see AM as fueling a new kind of consumerist frenzy where "hobbyists make legions of *white elephants* out of toxic plastics and [. . .] landfills are chock-a-block with yesterday's badly made fashionable shapes" [2].

2.4.1 Yet Another Industrial Revolution

A recurring claim is often made of how new paradigms of manufacturing will occur, often under the umbrella term of a "3rd Industrial Revolution" [1, 3]. The Industrial Revolutions of the eighteenth and nineteenth centuries had drastic social, economic, and political consequences for the lives of those living in developed countries. The improved use of water and steam power, combined with the development of new machine tools, made it possible to materialize new artifacts on an unprecedented scale. Whereas low-volume, decentralized craft production was the previous norm, the industrial rise brought with it centralized clusters of high-volume manufacturers. Starting off with the production of textiles, the notion of economies-of-scale [23] became the mantra for which every aspect of industrial development stood by.

An important aspect of economies-of-scale is the operational optimization of the factory floor. The production of technological artifacts, such as Henry Ford's T-Ford car, required the production and assembly of several thousand parts. Dividing the assembly of the T-Ford along a production line ensured that Ford's cars could be sold within an obtainable price range for millions of people, as long as they chose the color black. This concept of division of labor, with unskilled laborers performing repetitive tasks on the factory floor, strives to minimize the lead time and facilitate high-volume distribution, always aimed at minimizing the sales cost of items. While in modern production environments the methods and processes of material extraction, energy use, and production are being continuously refined, the underlying principle remains the same. Modern companies that can deliver the highest quality for the lowest price have the competitive advantage [32].

The envisioned forthcoming Industrial Revolution, fueled by the potentialities of digital fabrication, is often summarized as a shift from mass production to individual production. The concept of economies-of-scale, where similar plastic artifacts are made affordable due to the gradual down payment of expensive tooling, is complemented by economies-of-one. Here, unique, personal, and individualized artifacts function as the source of competitive advantage. This managerial concept of mass customization [34] focuses on a shift from the offering of generalized mass market products and services to the tailoring of solutions to specific needs, the rationale being that every customer has specific needs that cannot be addressed in a generalized way. Such a shift will necessarily influence many operational aspects of the manufacturing process, from a supply chain that will rely on nonlinear local collaboration to distribution, which will necessitate a direct communication between producer and consumer, and new business models that capture, create, and deliver new values [37]. As to the consequence of emerging and disappearing of professions as a result of increased robotization of labor, Lipson and Kurman (2013) draw similarities with how the Internet made many travel agents obsolete, while at the same time offering new possibilities for travel-related services [25].

2.4.2 Toward Economies-of-One

Because of the potentials for local manufacturing at a reasonable cost (depending on the product) and newfound design possibilities for personalization, AM can be tightly bound with online, customizable services. These were important motivations for making the website fabrikkaho.no (Fig. 2.5), an online sales platform featuring cell phone casings designed and manufactured by design students at the Oslo School of Architecture and Design (AHO). The concept, titled Fabrikk AHO (translates as Factory AHO in English), presents 21 individual design projects that are designed specifically for AM. The design in Fig. 2.5 shows a smartphone casing featuring an integrated kickstand and cardholder that allows for user customization of both texture and a 10-character phrase. While serving as a visual example of how online customization can be facilitated, the idea of economies-of-one can be entwined in product service offerings to various extents.

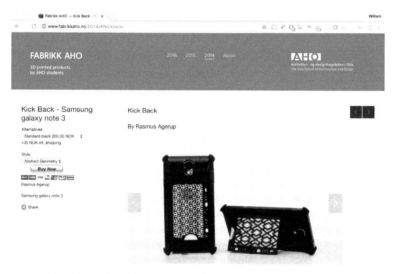

Figure 2.5 A 3D-printed kickstand for the Samsung Galaxy Note 3 cell phone, with customizable patterns, sold as a limited-edition product by design students at www.fabrikkaho.no. Design: Rasmus Agerup.

Later in this book Monika Hestad analyzes the role of AM in developing the eyewear brand Mykita, through a product and branding framework. As a part of her analysis she discusses both the internal and external drivers for Mykita as a brand. Mykita, which was founded in 2004, can be seen as a relative newcomer to the eyewear industry, mainly dominated by major fashion houses. Mykita presents itself as a modern eyewear company that combines modern production technology with traditional craftsmanship.

Under the banner of "Handmade in Berlin" its Mylon series brands itself as being both technologically innovative and sporty, honest, and imperfect. Produced in-house with powder-based SLS 3D printers, the Mylon series, which was initially developed in 2007, doesn't offer any individual user adaptation but, instead, offers a wide spectrum of variations, currently consisting of 51 different styles.

The company's website (mykita.com) presents Mylon not only as a product series but also as a complex material composite that is prepared in a series of stages. Starting off with the fine-powdered polyamide powder commonly used in SLS printing, the laser-sintered artifact is then cleaned, sanded down, color-dyed, and left to cure. It is conceivable to think that Mykita would have had the initial idea of making user-customizable glasses when the initial 3D-printed glasses, Mylon, were conceptualized. For Mykita, this would bring large implications for both its supply chain, as no two glasses are the same, as well as the customer journey. Where should the customer have his or her face scanned? How could the glasses be adjusted not only ergonomically but also as per the customers' individual preferences?

Almost 10 years after the development of Mylon, Mykita announced in 2016 its entry into the area of customizable eyewear through its MyVeryOwn series, which extends its material composite to incorporate individual ergonomic considerations. Still, the user journey for Mykita's personalized glasses starts with the physical shop, as shown in Fig. 2.6. The recent proliferation of 3D printing creates an innovative surge in technologies that supplement it. Consequently, new businesses and entrepreneurs fill in the need for specialist services and technologies, such as 3D facial scanners and optimization algorithms, that only recently have made MyVeryOwn possible. In Chapter 4, Hestad gives further insight into the branding of AM through the Mylon case.

How does a business perspective shape our understanding of the 3D printer in relation to design?

While the inventors of 3D printers removed some of the traditional barriers related to the fabrication of goods, their role as a manufacturing tool requires specific strategies for integration. The social group of business managers applies managerial concepts, such as economies-of-one, in order to leverage the potentialities of 3D printing into its value creation process.

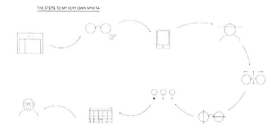
THE STEPS TO MY VERY OWN MYKITA

Figure 2.6 The figure shows how Mykita envisions its new services, focusing on personalization of its MyVeryOwn subbrand. Much like its currently available products, the first step on the user journey is found in the shop. Image: Mykita.

In the following section, these potentialities are further unpacked, through the perspective of design practice.

2.5 Designers' Perspectives of 3D Printing Futures

From the perspective of design, 3D printing has traditionally been used as a developmental tool for making VDRs, such as prototypes or mechanical verifications of to-be artifacts. Early adopters of RP, such as General Motors and Electrolux, rationalized the use of RP as a way to decrease time-to-market for new products. Early 3D printing technologies became important features in their corporate development strategies. Not only did this apply to the initial "fuzzy front end" [21] stages of a product development process, it also served a purpose for the marketing of new products, through beta testing, and in making visionary concepts.

Figure 2.7 shows the conceptualization of a computer mouse whose main components include a tactile, wooden surface mounted on a principle SLS-printed structure. Tasked with the challenge of conceptualizing a digitally fabricated computer mouse, the designer chose wood as the tactile material toward the user's palm. As opposed to carving the wood by hand, or using a computer numerical control (CNC) mill to subtract the form from a solid piece of wood, the designer created several rapid iterations of a 3D-printed mold using a desktop 3D printer. The mold was then placed in a vacuum-forming machine and used to bond several layers of wooden veneer together with a plastic polyethylene terephthalate (PET) (commonly found in soda bottles) sheet acting as the female part of the mold. Although the mold was rendered unusable after the initial trial, due to the low melting temperature of the PLA plastic used by the desktop printer, it was sufficient for the conceptualization of a novel manufacturing technique of a 3D-printed computer mouse. This example shows the dynamic use of fabrication tools, both traditional and digital, by designers who conceptualize new products and services with 3D printers.

Figure 2.7 A digitally fabricated computer mouse using several production techniques. Design: Hans-Martin Erlandsen; photo: William Kempton.

The student's work corresponds to a categorization of the use of 3D printing from the perspective of product design, which can be placed in three separate categories of actual developmental product design work [20]. The primary, and traditional, use of 3D printers lies in their ability to create design representations, such as visual models or quantitative structures of a physical design. Secondly, the use can be defined through its role in a manufacturing process, such as the making of the mold used for veneering of the computer mouse in Fig. 2.7. Thirdly, the use of a 3D printer in a developmental process can be defined as manufactured objects, where it acts as a production platform of end-user artifacts. This last categorization is often described as AM in the literature as well as in popular media.

2.5.1 Design and Additive Manufacturing

The shift in focus, from seeing 3D printers as developmental prototyping tools from an RP perspective to considering them as a means of production from an AM perspective, is a transformative shift. While designers and engineers were early adopters of 3D printing for making prototypes, new users of 3D printing are pushing forward an integrated development for conceptualizing, designing, and producing innovative new products. The beneficial abilities of AM production have been the basis for tentatively grouping [19] the potentials of AM as enhanced design, custom design, and computational design.

Both *custom* and *enhanced* can be seen as ways of describing a design or making use of it in the development of a product. Shortly described, a custom design may refer to a design where the dimension and style are based on unique, individual preferences. Enhanced design may refer to the improved capabilities of the AM process, such as making complexity feasible, while computational design relates to a process of algorithmic computation. Envisioned as artifacts, these potentialities can be seen as coexisting, as in the example of the animalistic coffee cups seen in Fig. 2.8. Here, the aesthetics of the design concept is envisioned around a series of dog-like gestures.

Figure 2.8 3D-printed, customizable coffee cups in ceramic, based on the movements of an animal. Design: Izelin Tuulikki O. Tujunen; photo: Inger Steinnes.

2.5.2 Designing with Technology

As a developmental process, human-centered design (HCD) is pulled toward the creation of products and services that are feasible and create value for their users. At the same time, increasing demand for digitalization, connectivity, and systematization create a need for designers to adapt to new emerging technologies. This is seen in areas of both product design as well as HCI, where new technological platforms such as virtual reality (VR) environments, 3D printers, and mobile platforms invite for new novel uses and implementations into current contexts. More often than not, designers are challenged with satisfying the relationship between tackling the creation of meaning and technological innovation.

As canons of the discourses on HCD and innovation, Norman and Verganti (2012) analyze the relationship between technology and meaning in their discussions on radical innovation versus HCD [30]. They argue that the incremental nature of HCD is adapted to developing meaningful designs, while radical innovation pushes for technological innovation. Their theoretical frame sets technology and meaning as two dimensions of innovation, with the biggest changes in innovation coming about when both dimensions change. Such technology epiphanies come about when new contexts are facilitated through technological innovation.

Pioneering in design research in AM over the last two decades, Steinar Killi (2013) discusses the need for the product design discipline in looking beyond an isolated technological push, a view often adopted by the corporate AM industry, which is the traditional innovator of digital fabrication technology [19]. His recently developed approach (see Chapter 3) to product design and AM, labeled "AICE" (adapt, integrate, compensate, and elongate), outlines a way of designing meaning through technology as opposed to designing from technology. I argue that such a view is relevant both in the case of HCD development as well as for gaining a situated understanding of the emerging role of 3D printing technology. Such a view forces us to consider the deeper, underlying discussion of technology in relation to design.

2.5.3   An Undetermined View of Design

In discussing philosophical approaches to technology, Ihde uses the notion of designer fallacy to explain the notion of designing intents with technology. Ihde calls for an unpacking of the complex relations that over time are played out between designers, technological entities, and the end users of technologies. The notions of intentional fallacy in literary theory argues that a text can be established only after the author's intentions are uncovered. Similarly, Ihde develops an argument based on the concept of designers employing purpose and use in technology, which is critiqued by Ihde. This notion that intent is not possible to determine fits into Ihde's undeterministic ontological view of technology and society. Such an undeterministic view is useful for our discussion of design, as it allows us to consider the technological materialities, qualities, and possibilities of 3D printing in a way that is useful to our view of design as a situated, open-ended process.

Such an open-ended view of 3D printing is similarly taken up by Peter Troxler [41] in his discussion of community-based fabrication laboratories (fab labs) as users and developers of 3D printing technology. Troxler also discusses the emerging awareness of value creation, but from the point of view of community engagement. His discussion of technology takes on an attitude of "you don't own it if you can't open it," arguing that technological empowerment is necessity for critical use of technology.

From an HCI point of view of technology and digital fabrication, Matt Ratto takes a similar stance in his use of freely available software for materializing digital information [35, 36] in his 3D printability project, which seeks out to make below-the-knee prosthetics.

How does a design perspective shape our understanding of the 3D printer in relation to design?

In developmental design practices, the use of 3D printing tools can be placed within three categories: as design representations, as a toolmaker for molds and fixtures in a manufacturing process, and as manufactured end-user objects. The latter category, which can be described as AM, comes with its own set of potentialities, which are tentatively customizable, enhanced, and computational.

However, in a wider scope of design practice, which is found outside of the confines of consumer-oriented practices, 3D printing can be seen as a tool that enables and empowers design engagement in new contexts and uses. The following section, which focuses on the social group of layperson makers, can be seen as one of the new contexts.

2.6 A Layperson's Perspective of 3D Printing Futures

As opposed to expert practitioners who are proficient in their given profession, a layperson can be described as a person without any specialist knowledge. With desktop fabrication becoming increasingly available in schools, in libraries, at home, and in community spaces, such as fab labs [43], it opens up opportunities for layperson participation in both formal and informal material making. In the section on the 3D printer inventors, I discussed how the recent developments in desktop 3D printing have strong ties with maker spaces. These informal meeting grounds are places where companies such as Ultimaker and MakerBot were founded and where initial production of their tools was taking place. In addition, these arenas serve as a playground for an increasing number of enthusiasts who make and share ideas with each other. Laypersons' involvement in 3D printing, therefore, forms a relevant social group in my analysis.

2.6.1 A Layperson as a Maker

Layperson involvement in 3D printing is often closely linked with the umbrella term "the Maker Movement." Popular magazines, such as *Make* magazine, review and discuss the evolving market of desktop 3D printers, in addition to organizing popular festival concepts such as Makerfaires. Independent hackerspaces also help mediate the use of digital fabrication tools as personal, desktop-friendly tools. On reflecting on the emergence of the Maker Movement, Dale Dougherty (2012), the founder of *Make* magazine and Makerfaire, emphasizes a return to a material engagement that makes people more than just consumers. Dougherty describes a "maker" as a holistic ideal—"We all are makers: as cooks preparing food for our families, as gardeners, as knitters" [9].

Since the release of certain 3D printing patents [6] the availability of desktop 3D printers has gradually increased among nonexperts. From the initial self-replicating open-source RepRap tools to preassembled tools such as Ultimaker, 3D printing tools and commodities are becoming increasingly accessible for layperson use. 3Dhubs.com, a social distribution platform for 3D printing services, notes that user-to-user 3D printing services are now available in every continent, with cities such as Milano and Amsterdam hosting up to 300 individual hubs each. While local users are supplying laypersons in many of these European cities with prototypes and models, access to local expertise is also happening through initiatives such as maker festivals and in local maker spaces.

2.6.2 Making in a Learning Environment

The image in Fig. 2.9 shows a group of high-school students assembling an Ultimaker 3D printer at a local maker festival in Oslo. The kit was donated by local organizations and is distributed to several high schools in Oslo. As the printer contains a lot of moving parts, it is prone to jam. By assembling the kit themselves, the students get an insight into the inner workings of the technology, thereby making it easier for them to identify future problems.

Figure 2.9 Local high-school students are assembling Ultimaker 3D printers at a local maker festival in Oslo. Photo: William Kempton.

An increasing number of educational environments are adopting digital fabrication tools in their curriculum, as it is coming to represent the twenty-first-century equivalent of a shop class. Organizations such as MakerEd, which are developing resources and online libraries through the vision of "Every child a Maker" [26], are some of the many organization that see 3D printing as a powerful educational tool. Although the use of digital fabrication can be seen as relevant for its engagement in both math and engineering, it is important to consider that engagement in 3D printing doesn't necessarily need to relate to any specific competencies. As we are arguably in the "early phase of a wide-scale revolution in tangible creation" [10], the adoption of digital fabrication in the day-to-day culture of children and youth can have positive educative traits.

Performing tasks, failing, and redoing them as a reflective process reflects a mode of learning that emphasizes experience.

As an influential educational reformer, John Dewey [8] argued that there are different forms of experiences, agreeable and disagreeable, that naturally affect future experiences [8]. Dewey criticized the contemporary education of his time in that it built on various established routines for teaching existing knowledge, which would later be imposed on the younger generations. Instead, he proposed a mode of experiential learning that is linked to real-world objects and not bound by the model of current natural science (STEM) education way of organizing subject matter [8]. This ideal of an experiential learning process was later formalized by Kolb [22], who supplemented the idea of figurative representation of experiences with a transformation of that representation [22]. In such a way, the theory of experiential learning relies as much on making as it does on reflecting.

Digital fabrication tools have the ability to facilitate an experiential learning cycle. The 3D printer is in fact a tool for making physical, real-world artifacts. And as with most making processes, its very nature is incremental. An idea is deceived, deliberated upon, and conceptualized. However, the concept of 3D printing requires extensive knowledge in a variety of fields (although the inventors of desktop printers would claim otherwise). Not only does the 3D printer rely on a series of subtools, such as apps for preparing content to be fabricated, it also relies on a material input. As the 3D printer is a tool for fabricating real-world artifacts it requires physical ingredients, usually in the form of plastic, and also a digital blueprint.

How does a layperson perspective shape our understanding of the 3D printer in relation to design?

Layperson engagement in 3D printing is an increasingly relevant topic for design. It relates to discussions on consumerism, education, and distribution and production of goods and artifacts. It can also be seen as a tool for opening up the notion of making as a form of literacy and the ability to question the consumerist patterns that we live by. These are some of the central ideals of the Maker Movement.

2.7 Discussions and Conclusions

2.7.1 Summarizing the Perspectives

In summarizing the various perspectives presented in this study, digital fabrication tools can be seen by the first generation of 3D printer inventors as engineering-oriented tools, facilitating the manufacture of other things, through prototypes and as VDRs. While RP remained the dominant purpose, there also existed early visions of making consumer artifacts directly with the 3D printer. However, development of the desktop-oriented 3D printer by the second wave of 3D printer inventors somewhat realized this concept, targeted toward tinkerers and hobbyist makers.

The social group of business managers strives to create new value systems within a paradigm of economies-of-one, with key problems seen as the necessary operational restructuring of production and delivery.

The perspective of the social group of designers is more entangled in using 3D printing as a developmental design tool. The developmental use of 3D printing can be placed within three categories: as design representations, for making tools such as molds and fixtures in a manufacturing process, and as manufactured end-use objects, often referred to as AM. In the periphery of consumer-oriented design practices, 3D printing is increasingly explored due to its novel material, aesthetical, and empowering potentials.

The technological frame of laypersons, often referred to as layperson makers, is concerned with conceptualizing and making for private consumption. However, the open sourcing of 3D printing techniques is increasingly blurring the lines between the 3D printer inventors and laypersons. Laypersons are challenged with acquiring adequate proficiencies for material making, as well as having accessibility to 3D printing tools.

2.7.2 3D Printing Futures

The story of the development of the 3D printer is by no accounts written. In an interpretive analysis such as this, there can be several analyses that forward an argument of sociotechnical development.

In any such analysis, the matter of choosing which relevant social groups to discuss is equally important. For this chapter I have selected a set of social groups that allow me to analyze 3D printing not only as an instrumental tool for experts but also as an increasingly democratized tool for making. Through my analysis of recent social and technical events I attempt to build an understanding for evaluating contemporary as well as future understanding of digital fabrication. And as much as my analysis of the development of 3D printing up to now is portrayed in the light of my selection of relevant social groups, the future visions of 3D printing are increasingly up for interpretation.

As there is a continuing amount of interest in the future role of digital fabrication, so is the continuing amount of interested parties. The second wave of 3D printer inventors, such as Ultimaker and MakerBot, developed and made their first fabrication tools out of maker spaces, primarily serving a clientele of self-proclaimed makers. While newly founded companies such as these continue to grow, the next wave of 3D printer inventors are of a different sort. Large technology companies not previously engaged in 3D printing, such as HP, recently stated in a press release that their future growth now lies in digital fabrication. This was later supported with the release of a newly developed fabrication technology, HP Multi Jet Fusion, which builds on much of their knowledge and expertise from traditional inkjet printing. While HP develops the fabrication process, it closely collaborates with the chemical industry for developing new materials. Companies such as BASF (Badische Anilin und Soda Fabrik), a major chemical and plastics manufacturer, might have a considerable impact on the employment of AM in future contexts. As they improve and create new materials for AM, they make themselves relevant as far as a social group goes.

2.7.3 Constructing a View of Sociotechnical Development

Through an argument that is based on a social constructivist idea of technological development as being socially constructed, this book chapter analyzes the emerging social practice, mediation, and knowledge being brought through the development of the 3D printer. These theoretical frameworks that have been applied are brought in from critical theories of STS, such as those by Feenberg, Ihde, and Bijker, which argue for a view on technological development as being undetermined.

The theoretical concept of relevant social groups and technological frame is introduced through the story of the development of home video entertainment. Similarly, the development of the 3D printer can be seen in the light of its relevant social actors. For my analysis I have focused on the relevant social groups as seen from a design-relevant perspective. The 3D printer being a tool that facilitates the making of things, I argue that it has great implications for the way design is being conducted, as well as its emerging role as a tool for democratizing design for the current and future uses of the digital fabrication tool.

Appendix: Technologies for 3D Printing

The 3D printer is a technological device for additively fabricating curated, sensory content. Like other fabrication devices, such as sewing machines and inkjet printers, the 3D printer delivers its medium in the form of physical artifacts that can be touched and felt. Whereas an inkjet printer prints text and images on a 2D surface, the 3D printer fabricates 3D artifacts according to a set of predefined instructions. As a technological principle, 2D printing can be divided into many different subprinciples along a historical time line—from woodblock printing, which dates back to early Asian culture, to the Gutenberg press, which initiated the European age of printing. More recent innovations include the introduction of photocopying by Xerox in the early 1960s to inkjet and laser printers, which gave way to the concept of desktop publishing.

Similarly, 3D printing techniques make up a wide-ranging set of tools, as the formats, or processes, have distinct properties (Fig. A.1). Some processes are unique in the way they reproduce large, transparent artifacts, while others create small, highly detailed objects. One could also speculate that the reason for the coexistence of all the different 3D printing processes is that these devices have only recently become domesticated. Only after certain patents for digital fabrication processes expired, namely the patent for filament deposition modeling printing (Fig. A.2c) developed by Scott Crump [2], did the desktop 3D printer become available as a domestic appliance. Starting with the open-source development of the self-replicating RepRap printer [1], the market for relatively low-cost filament deposition modeling printers has proliferated. Companies such as Ultimaker, WASP, Printrbot, and Zortrax, to name a few, sold printers by the thousands to enthusiasts all over the world.

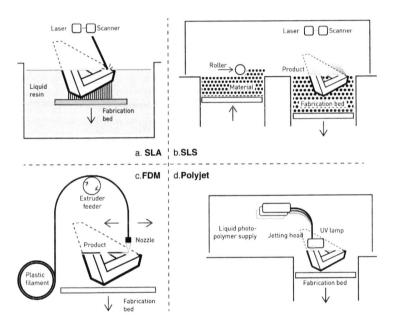

Figure A.1 Some of the processes that make up 3D printing: (a) Stereolithography (SLA) involves the use of a photopolymeric liquid, which is selectively hardened by exposure to light. (b) Selective laser sintering (SLS) deposits thin layers of power-based plastic, such as nylon, which is heated up and sintered together by a powerful laser. (c) Filament deposition modeling feeds strings of plastic filament through a nozzle, which gradually builds up the artifact. (d) PolyJet, similar to the concept of SLA (which uses a photopolymeric liquid as a principle building material) deposits and selectively hardens voxels, 3D pixels of material, in thin layers. Illustrations: William Kempton.

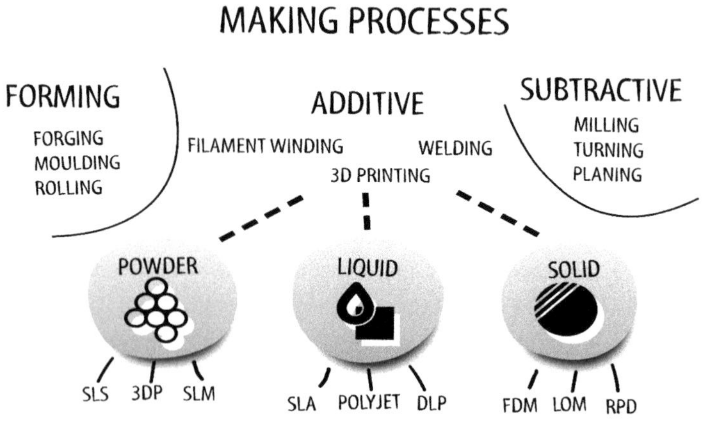

MAKING PROCESSES

Figure A.2 The processes for making can be described as being either forming, additive, or subtractive. 3D printing is an additive process as it is operates by binding or bonding material together in a layerwise fashion. Illustrations: William Kempton.

A.1   The Principle of 3D Printing

Much the same way as you would bake a cake, the principles of 3D printing usually involve a recipe, a set of ingredients, and a mechanism for bonding, curing, or sticking them together. But unlike the manual fashion in which you would compose the cake, the 3D printer eliminates the need for intervention by automating the forming process. The different technologies that make up 3D printing require both different recipes and ingredients, as the processes differ. Some processes, such as filament deposition modeling, apply strings of plastic polymer in a layerwise fashion, while others, like SLA, rely on a liquid that reacts to light. There are powder-based machines that spread layer upon layer of fine-grained plastic, gypsum, or metal powders onto a surface, which are then bonded together. Sheets of paper may also be cut, glued, and stacked together to form a 3D object.

Current powder-based processes create precise and structurally strong objects with a high degree of design freedom. Liquid-based processes imitate well-known materials and materialities (such as transparency, translucency, and flexibility). Lastly, solid-based processes such as desktop 3D printers are low-cost technologies often used for prototyping. Figure A.3 illustrates how the many making processes of 3D printing can be described as being powder based, liquid based, and solid based. What they all have in common is that they are *additive* forming technologies.

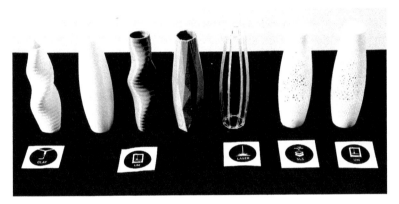

Figure A.3 A series of cylindrical shapes that are made using various digital fabrication techniques. Models: Christopher Pearsell-Ross; photo: William Kempton.

The recipe for which the 3D printer gets its instructions is a crucial part of the digital fabrication process. Much the same way a craftsperson, chef, or laborer relies on a set of instructions to make a product, the 3D printer relies on a set of inputs to perform. The 3D printer is in fact a "CAD-based automated additive technology" [3] as it relies on a digital blueprint typically from a computer-assisted drawing (CAD) program. These drawing tools were once restricted to "experts" as they were expensive and required a steep learning curve. However, as the popularity of 3D printing has increased, so has the availability of CAD programs to nonexpert users. New products and services that focus on nonexperts are being promoted from companies such as Autodesk, Solidworks, and Onshape. This new generation of CAD programs is becoming more accessible through new pricing schemes, online tutorials, and the fact that they are cloud based. This again opens up for use on mobile devices, such as smartphones and tablets.

A.2 Additive Manufacturing as a Making Process

Natural phenomena that create rock formations or allow trees to grow are natural making processes that construct and continuously evolve the earth's surface. Similarly, the act of making objects and artifacts through tools and processes is an important aspect of human endeavor. But as opposed to natural processes, such as photosynthesis, ocean currents, and seismic activities, tools made by people are constructed, made artifacts. As they are put to use, as part of a process, the artifacts become part of a technological system. In essence, the objects and artifacts that we surround ourselves with every day are all products of technological making processes. And as with nature, these methods of making can roughly be described as *subtractive*, *forming*, or *additive* processes (Fig. A.2). As erosion chips away at soil and rock, it forms new land and scenarios. Similarly, man-made *subtractive* technologies rely on solid pieces of material such as wood, stone, or foam to be cut, milled, planed, or trimmed away. And whereas these technologies once relied on manual labor, such as a craftsman's chisel, digital fabrication tools such as computer numerical control (CNC) milling have become its modern extension.

As shells are bones are left in the ground for millions of years, the surrounding sand and soil leave fossilized imprints of what was once a living organism. As the organic material is replaced by minerals inside the cavity of the shape, nature manufactures an internal mold of itself. Similarly, man-made processes for molding, forging, rolling, or deforming material into new form can be characterized as a *forming* process. In the manufacture of plastic parts, a tool, containing the hollow cavity, is used to reproduce exact, positive replications of itself.

As for *additive* processes, these are also abundant in nature, for example, trees and plants that grow and wounds that heal. However, in the manufacture of objects and artifacts, the process is relatively novel. As an additive process, 3D printing is capable of transforming material into new purpose, without the need for a predefined tool or mold determining the shape of the artifact being made. A coffee cup that has been slip-casted in a mold may be identically reproduced thousands of times from the same mold. A coffee cup made by additive means, however, may be reproduced thousands of times, each with a unique shape. There are many potentialities of additive processes that go beyond aesthetic freedom, such as material composition and potentially less material consumption. Some even claim 3D printing has the potential to greatly democratize the acts of both designing and manufacturing things [46].

In the following sections we further explain the various additive manufacturing technologies. These can be seen as powder based, liquid based, and solid based.

A.3 Powder-Based Additive Processes

A.3.1 Selective Laser Sintering

A.3.1.1 SLS background and process

SLS (Fig. A.4) has as one of the principal methods of rapid prototyping a history of producing prototypes and tools for other manufacturing methods, such as injection molding. The technology, first commercialized by Carl Deckard in the 1980s, is currently one of the most commonly utilized technologies for making both prototypes and end-user parts (additive manufacturing).

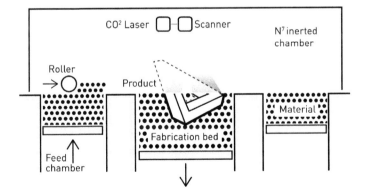

Figure A.4   SLS process principle. A powerful laser sinters particles of material together, layer by layer.

In preparation of the fabrication process, a large amount of the material is filled into two separate feed chambers, one on each side of the fabrication bed. The SLS build chamber is then filled with nitrogen-rich air and heated to approximately 170°C (depending on the material), while a base of material, usually polyamide, is spread out by the feeder mechanism (while 3D Systems employs a roller, EOS uses a wiper due to patent disputes), bringing material from the feed chambers at each bypass. The fabrication bed is held just below the material melting temperature, before a 12°C increase caused by the passing laser sinters the material granulates together. New layers of material are fed over each other and repeatedly bonded. The build process, depending on the load, fabricates approximately 15 mm of material an hour.

After the fabrication process the entire fabrication build is set to cool down to below 70°C, before the fabrication "cake" is lifted out to a cleaning station. The excess nonsintered material is often partially reused. The fabricated artifact is then blasted with abrasive glass powder to clean out the remaining powder.

A.3.1.2 Design considerations for SLS printing

The ability to fabricate strong isotropic parts, in addition to the nonsintered powder acting as a support for cantilevered parts, makes it possible to create a wide variety of complex shapes. It is well suited to fabricate open-lattice structures and perforated surfaces as both large solid geometries as well as flat planes may suffer deformation. As the material is self-supported, free-hanging shapes can be printed inside a cavity. However, if the cavity is closed, material will be trapped in the cavity. An important design consideration is to create holes for nonsintered material to exit. Common to all the powder-based processes is that the surface finish will be matte and porous and slightly water permeable. Postprocessing of the fabricated artifact is common, from spray painting to dyeing or polishing.

A.3.1.3 Materials in SLS processes

A common material used in the SLS process is white PA-12 polyamide (PA) powder, which gives strong yet flexible parts with a high material finish. Other materials, such as glass-filled or carbon-filled PA, provide unique material capabilities such as increased material stiffness and thermal conductivity, to mention a few. It is also possible to fabricate using flexible, elastomeric materials, as well as HD-PE, PET, polystyrene, PA-11, and PA-6.

A.3.2 Inkjet Powder 3D Printing

A.3.2.1 3D printing background and process

3D printing (Fig. A.5) is a relative newcomer to the market of 3D printers, developed by researchers at MIT in the early 1990s and made available in 1993. The ZPrinters, as they were initially called, operate in a layerwise fashion, selectively fusing cross sections of material much like any other 3D printing processes. Instead of using heat to fuse the cross sections, which are filled with gypsum powder, an inkjet printer head moves around depositing droplets of binder liquid, thereby fusing the gypsum. The process makes it possible to fabricate mock-ups and models in greater speed compared to other technologies, at lower cost, and with the ability to fabricate colored artifacts.

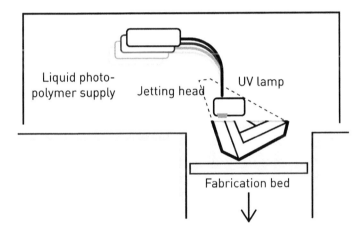

Figure A.5 3D printing deposits a liquid adhesive onto a layer of gypsum powder through an inkjet print head.

The use of plaster-based materials, in combination with a binder material that reacts with the plaster, is unique to 3D printing technology. While successively depositing layer upon layer of fine-grained plaster powder, a set of color print heads moves and selectively deposits the binder, similar to an inkjet printer. Whereas an inkjet printer prints on paper, the print heads in the 3D printing machine print directly onto the gypsum surface before a roller spreads another layer of powder from a feeder. The amount of binder that is deposited onto the surface varies: On a dense shape, the outer shell is fully saturated with both color and transparent binder, while the core of the shape is partially filled with binder. Artifacts are fabricated with relatively low temperatures. Although there is no heat fusion, the binder liquid causes the plaster to harden and radiate heat. After processing, artifacts are usually left in the machine's build chamber to dry and cool down. Once taken out of the fabrication chamber, the artifact is carefully brushed clean of material residue.

A.3.2.2 Design considerations of 3D printing

3D printing technology is often used for making prototypes and models. The plaster material used in the process is fragile, brittle, and slightly porous. The material is therefore well suited to grinding and polishing. The relative fragility of the material makes it unfit for functional or mechanical prototypes. When constructing for 3D printing, a wall thickness of less than 2 mm and details of less than 1 mm should be avoided. As the part is fragile during unloading, any unsupported walls or cantilevered parts should be at least 3 mm thick. Part orientation in the fabrication bed may affect tolerances and should be considered in the process.

A.3.2.3 Materials in the 3D printing process

While there are few officially compatible materials that work with 3D printers, third-party suppliers have developed ceramic powders with accompanying binders, which make parts that can be burned in a kiln. Although the plaster is fragile when taken out of the fabrication chamber, resins or mineral salts may be infused onto the surface for increased strength, saturation of color, or smoothness of surface.

A.3.3 Selective Laser Melting

A.3.3.1 SLM background and process

The selective laser melting (SLM) process (Fig. A.6) was initially developed by the German Fraunhofer Institute and became commercially available in the early 2000s. SLM is characterized by its ability to make high-value, low-volume, end-user parts from an increasing inventory of metals and alloys. The fabrication process has been adopted for additive manufacturing purposes by specialized industries, such as medical orthopedics and aerospace industries.

A fine-grained metal powder is spread across the fabrication bed, while cross sections of the 3D model are selectively fused together in a layerwise fashion. As opposed to other processes for metal fabrication, such as direct metal laser sintering (DMLS), SLM fully melts the powdered metal into a homogeneous mass using a powerful laser with several hundred watts. As the process is time intensive and therefore costly, it is considered a manufacturing process of end-user parts or for tooling, as opposed to making design representations.

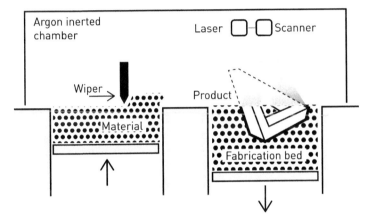

Figure A.6 Fine particles of titanium, steel, and other metals are spread over a thin surface before being exposed to a high-power laser, which welds the particles together.

A.3.3.2 Design considerations of SLM printing

As opposed to plastic powder-based processes, fabrication with metal powders may rely on additional support structures if the surface is less than 45°. Also, surface finish is best if the surface is fabricated in an upright position [9]. Support structures may require additional postprocessing.

Parts made in SLM may be of varying accuracy, as overhangs and holes may be subject to material curl, a condition where the top layer of melted metal powder lifts up from the fabrication bed. When designing for SLM, details of less than 0.3 mm and a wall thickness of less than 0.5 mm should be avoided.

A.3.3.3   Material in the SLM process

Apart from being powdered, the materials that can be processed with the SLM process need certain flow characteristics. Commonly used materials include stainless steel, titanium, aluminium, cobalt chrome, and tool steel.

A.4   Liquid-Based Additive Processes

A.4.1   Stereolithography

A.4.1.1   SLA background and process

Stereolithography (SLA) was first patented by Chuck Hull in 1986 and made into a working machine in 1987 and is arguably the founding principle for 3D printing. As with SLS, the technology was initially developed for rapid prototyping, making highly detailed visual models for the development process of products. Fabricated parts were brittle and vulnerable to light exposure, limiting their functional use. However, due to recent developments in material and process, SLA has become an important tool in the shift toward additive manufacturing.

The liquid photopolymer used in the fabrication process is sensitive to light and can be highly toxic, requiring special care when preparing the process. Upon contact with a UV laser, a thin layer of liquid is solidified to the fabrication bed (Fig. A.7). Although there are several principles for the fabrication process, they all have in common that the UV laser cures a complete layer before the fabrication bed moves in preparation for the next layer to be fabricated. The object either emerges from the vat of liquid or is gradually submerged into it. Between the curing of each layer, surface tension between the liquid and the solidified polymer is broken by either tilting the vat or sweeping over the most recently fabricated layer with a paddle. Either way, the object that is being made requires a set of support structures that make the parts stick to the fabrication bed, as well as allowing for overhangs to be constructed.

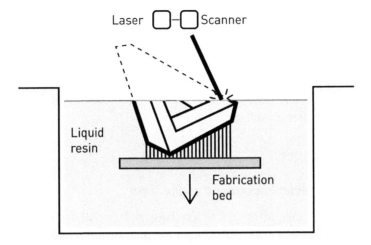

Figure A.7 A photopolymeric liquid is exposed to the light of a laser, which solidifies the liquid.

When the process is completed, the fabrication bed is removed, with the fabricated object stuck onto it, before it is broken loose and cleaned with alcohol-based chemicals. Depending on the material choice, the part is either posttreated with other chemicals or further cured under intensive UV light for a short period.

A.4.1.2 Design considerations of SLA printing

With layer thicknesses varying from 0.025 mm to 0.2 mm, the SLA printer is considered a process capable of producing high-resolution parts, capable of producing surfaces of near-injection-molded quality. The process requires additional support structures, which may require postprocessing. To minimize the amount of support structures needed, parts may be either tilted to minimize overhangs or specifically designed with angled overhangs. A common rule-of-thumb is to avoid overhangs of less than 30°.

A.4.1.3 Materials in the SLA process

The liquid-based materials used in the SLA processes are complex chemical compounds tailored specifically for each SLA printer. While both solid- and powder-based materials are based on common thermoplastic polymers such as PA, PLA, or ABS, the base materials used in SLA are often described in relation to their characteristics.

Materials are often described as ABS-like, flexible, castable, tough, or extreme. Clear materials, which are a unique characteristic of liquid-based processes, may be used to create translucent or near-transparent parts with varying thicknesses. While SLA materials have traditionally been seen as fragile, toxic, or vulnerable to light exposure, recent material developments are gradually rendering these limitations obsolete.

A.4.2   PolyJet

A.4.2.1   PolyJet background and process

In 2000, Objet Geometries, now a subsidiary of the 3D printing giant Stratasys, launched its first PolyJet 3D printer. Since then PolyJet and other multijetting technologies are available from several companies, such as 3D Systems. By selectively depositing droplets of photopolymeric materials, the process makes it possible to create multimaterial combinations with high resolution and accuracy. It is commonly used for rapid prototyping as it is able to reproduce and recreate several material qualities, from hard plastics to soft elastomers.

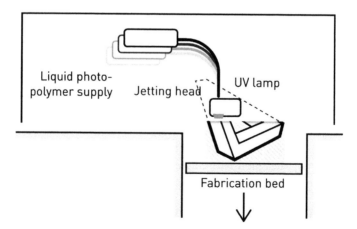

Figure A.8   After depositing droplets of a photopolymer through a print head, a UV lamp solidifies each layer of material.

Much like an inkjet printer deposits its inks onto a piece of paper, the PolyJet printer leaves droplets of photopolymeric liquids, which are then cured by UV light (Fig. A.8). Whereas ordinary paper printers produce pixels of color, the PolyJet printer produces 3D units of material, known as voxels. Similar to a filament deposition modeling printer, the PolyJet printer fabricates in a layerwise fashion in open space. Consequently, a certain amount of soluble support structure is deposited. Upon completion, the part is taken to a postprocessing chamber and the support is water-jetted away.

A.4.2.2 Design considerations of PolyJet printing

The PolyJet process is able to reproduce high-resolution surfaces, with layer heights down to 16 microns. This makes it possible to create models with a high degree of details, as well as having high mechanical properties.

A.4.2.3 Materials in the PolyJet process

PolyJet technology is unique for its ability to offer a wide range of materials, in combination with each other, or even with gradient properties. Available materials range from rigid general prototyping materials to rubber-like elastomers to ABS- or polypropylene-like materials to translucent plastics. Recent advances in material and process technology have also made it possible to print in a wide scale of colors.

A.4.3 SLA-DLP

A.4.3.1 SLA-DLP background and process

Similar to SLA, the stereolithography digital light processing (SLA-DLP) process uses a liquid photopolymer as the fabrication material. Whereas most SLA processes use a single laser source for curing the photopolymer liquid, SLA-DLP processes cure entire cross sections at a time using DLP projectors, similar to that of a home cinema. The principle was first used by Envisiontec in the early 2000s and has since evolved into technologies such as continuous layer interface production (CLIP), pioneered by Carbon 3D, which makes use of an oxygen-permeable layer between the resin tank and liquid. This enables the continuous curing of resin, which greatly speeds up the fabrication process.

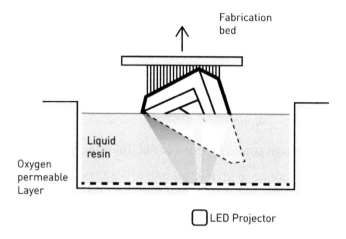

The principle, which cures entire cross sections of photopolymeric material at a time, has both advantages and disadvantages in comparison to traditional SLA techniques. Firstly, it has the potential to speed up the fabrication process, as the single DLP light source cures entire cross sections at a time. The DLP projectors may also generate gradient light sources, making it possible to cure several layers with different light intensities. However, SLA-DLP relies on a DLP projector as the light source, which has to be mounted under the resin tank and takes up considerably more space than a desktop SLA printer.

A.4.3.2 Design considerations of SLA-DLP printing

SLA-DLP printers are typically capable of fabricating with layer thicknesses varying from 0.015 mm to 0.15 mm. The SLA-DLP printer is considered a process capable of producing high-resolution parts and producing surfaces of near-injection-molded quality. The process often requires additional support structures, which may require postprocessing. To minimize the amount of support structures needed, parts may be either tilted to minimize overhangs or specifically designed with angled overhangs. A common rule-of-thumb is to avoid overhangs of less than 30°.

A.4.3.3 Materials in the SLA-DLP

Much like the SLA process, SLA-DLP materials are available in a range of material qualities, from clear, translucent plastics to biocompatible dental materials. Most of the available technologies make use of proprietary materials.

A.5   Solid-Based Processes

The simplicity of solid-based processes has in many ways become the public symbol of how 3D printers work. Found in schools, libraries, and workshop environments, the desktop 3D printer relies on a continuous string of material filament, usually thermoplastic, which is heated up by a moving extruder head and deposited in a layerwise fashion onto a flat surface. As each cross section of the desired artifact is constructed, the extruder head moves to the next layer and continues to deposit material. The process continues until the desired artifact is fabricated.

Common technologies that make up solid-based processes are fused deposition modeling (FDM) and fused filament fabrication (FFF), which all deposit material through an extruder. Similar principles, such as rapid plasma deposition (RPD), developed by Norsk Titanium, rely on additional subtractive forming methods to create precision details, as the deposited material consists of titanium in a plasma state. Other processes, such as Solidscape, deliver droplets of heated wax, which are then subtractively formed in order to create high-detailed positive shapes for lost-wax casting. While all the aforementioned processes involve smaller units of solid-based material being deposited, the laminated object manufacturing (LOM) process relies on sheets of material being cut and glued together.

A.5.1   Filament Deposition Modeling / Fused Filament Fabrication

A.5.1.1   FDM/FFF background and process

The concept of FDM was initially developed by Scott Crump, who was awarded the initial patent in 1992. In the years that followed, Stratasys turned the technology into a series of 3D printers, such as the Dimension and Fortus for industrial purposes. Soon after the key patent expired in 2009 [2], the marked for low-cost, open-source 3D printers proliferated as low-cost electronics and simple mechanical solutions made them feasible. The term "fused filament fabrication" (FFF) was coined by the RepRap project, whose goal is to openly develop a low-cost 3D printer that can print most of its own components [1].

Whereas industrial FDM processes use proprietary components such as material cartridges, dual extruders, and enclosed heating chambers, most of the FDM/FFF 3D printers are open structures made by off-the-shelf components. They both operate by extruding molten material through a heated nozzle, which is mounted onto a moving head. The extruder head usually moves on a flat plate (X and Y dimensions), depositing its material onto a fabrication bed. Upon the completion of each layer, the heated nozzle moves in relation to the fabrication bed (in the Z dimension), gradually forming cross sections of the artifact (Fig. A.9).

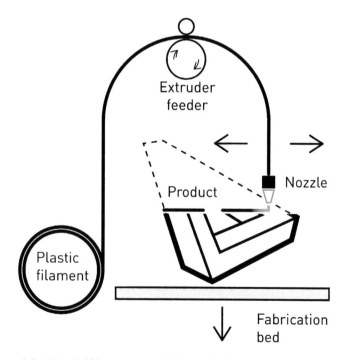

Figure A.9 Material filament extruded through a heated nozzle and deposited.

A.5.1.2 Design considerations of FDM/FFF printing

The FDM/FFF process operates in open air, which puts constraints on the design that is to be manufactured. A common rule-of-thumb is to avoid cantilevers that exceed 45° from the build plate when operating without support. Most industrial-grade printers utilize a dual-nozzle setup, with building material being extruded through one nozzle and a soluble support material through the other. Single-material, desktop-grade printers use the building material as support material, which has to be removed manually after the fabrication process. Free-floating enclosed parts are difficult to make as the process operates in open air.

As with most other fabrication processes, the FDM/FFF process operates in a layerwise fashion. Layer by layer of material is deposited with common layer thicknesses varying from 0.1 mm to 0.5 mm, depending on the nozzle diameter. The layers that make up the artifact are often visible, especially on near-horizontal sloped surfaces.

Depending on the material choice, fabricated objects are seldom isotropic; they are often significantly stronger in the direction of the fabrication bed. This is due to the fact that continuous strings are fused onto each other, with the bond between layers being significantly weaker, depending on the use of material. In some cases the fabricated object may even be water permeable.

A.5.1.3 Materials in the FDM/FFF process

Two convincing aspects of most solid-based processes are the low cost and wide range of materials available for FDM/FFF processes. ABS, a common thermoplastic, has traditionally been the material of choice as the material is well suited for extrusion. Many low-cost FFF printers prefer to operate with PLA, an organic thermoplastic, as the material allows for fabrication without the need for a heated chamber. Recent material developments have made it possible to fabricate with polymers such as PET, commonly found in soda bottles, as well as Nylon materials. A wide variety of blended materials such as PLA or PET mixed with bamboo, brass, copper, and carbon fiber are becoming popular among users of 3D desktop printers, because of the aesthetical and material qualities they offer. Conductive materials for capacitive touch applications are also available.

A.5.2   Laminated Object Fabrication

A.5.2.1   LOM background and process

LOM emerged as a method of fabricating prototypes using rolls of paper that were cut and glued in a layerwise fashion (Fig. A.10). The initial process, developed by Helisys, struggled with maintenance and technological issues [10]. The paper-based artifacts were best made as solid, thick-walled parts, while the industry was increasingly demanding prototyping tools for functional and detailed models. Although Helisys folded, more recent companies such as Mcor Technologies have met with success using the process. By using standard A4 sheets in combination with a desktop inkjet printer, models can be made both economically and in color.

Although the process principally uses rolls of paper, more recent technologies such as the Mcor Iris place layers of standard 80 gsm A4 sheets on top of each other, which are glued and cut, eventually forming the model. When making color models, each A4 sheet is preprinted with a color inkjet printer before being placed onto the fabrication bed.

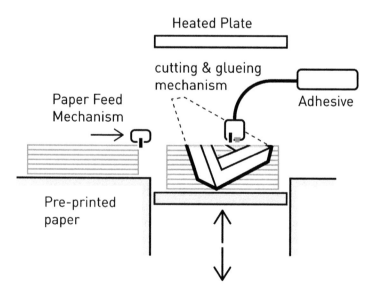

Figure A.10   Sheets of paper are cut, glued, and heated, layer by layer.

A.5.2.2 Design considerations of LOM printing

A minimum wall thickness of 4 mm should be considered, providing enough area for proper glue adhesion. Parts will have less strength in the Z direction relative to X and Y, although cohesive strength can be increased by applying glues or other adhesives.

A.5.2.3 Materials in the LOM process

As the principal material in use is paper, postprocessing, such as cutting and drilling, is easy. The LOM process is not typically used for functional parts.

A.5.3 Rapid Plasma Deposition

A.5.3.1 RPD background and process

Titanium alloys are traditionally known for being difficult to machine. In addition, 90% of the material may need to be cut away, making the forming of titanium both time consuming and costly. By combining both additive and subtractive forming principles, RPD is able to manufacture end-user parts from specialist metals such as aerospace-grade titanium. The technology, pioneered by Norsk Titanium, fabricates a near-net shape, which is up to 80% complete, before surface finishing and detailing with a CNC mill (Fig. A.11). The process uses both less energy and less time compared to conventional milling of billets. The technology is able to deposit 6 kg of titanium per hour and is aimed toward building larger structural components [6].

Using the principles of plasma arc heating, titanium wire is heated and deposited into a substrate in a cloud of argon gas. The near-net shape is fabricated in a layerwise fashion. Continuous process monitoring ensures that the layers consist of seamless, homogeneous structures. The deposition bead, which is approximately 8 mm in width, is then milled to provide a smooth surface.

A.5.3.2 Design considerations of RPD printing

The technology is developed for larger structural components, up to $900 \times 600 \times 300$ mm^3 in size, and is not optimized to produce small embossed or debossed details. Being a composite fabrication process, RPD can reproduce any shape rising vertically out from the substrate. The technology has no capacity for creating support structures. Surfaces inclining from the substrate are limited to 30°.

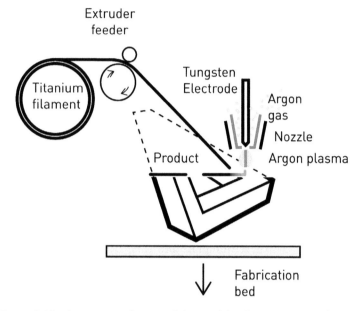

Figure A.11 A near-net shape is fabricated by fusing titanium filament through exposure to argon plasma before it is milled and surface finished with a CNC mill.

A.5.3.3 Materials in the RPD process

The technology is primarily developed around the use of titanium. The material exists in several alloys, Ti-6Al-4V being a common specification. It is known for its ability to be resistant to corrosion in seawater and chlorine. A low density and high strength ratio, fatigue resistance, resistance to heat creep, and biocompatibility make titanium sought after in as diverse circumstances as jet engines, prosthetics, medical implants, and bicycle frames.

References

1. Anderson, C. (2012). *Makers: The New Industrial Revolution.* New York: Random House.

2. Armstrong, R. (2014). 3D printing will destroy the world. Retrieved October 27, 2016, from https://www.architectural-review.com/today/3d-printing-will-destroy-the-world/8658346.article

3. Berman, B. (2012). 3-D printing: the new industrial revolution. *Business Horizons*, **55**(2), 155–162.

4. Bijker, W. (1997). *Of Bicycles, Bakelites, and Bulbs: Toward a Theory of Sociotechnical Change* (New ed.). Cambridge, MA: MIT Press.

5. Birtchnell, T., and Urry, J. (2016). *A New Industrial Future?: 3D Printing and the Reconfiguring of Production, Distribution, and Consumption.* New York: Routledge.

6. Crump, S. S. (1992). Apparatus and method for creating three-dimensional objects. Retrieved June 9, 1992, from http://www.google.com/patents/US5121329

7. Daly, A. (2016). *Socio-Legal Aspects of the 3D Printing Revolution.* Springer.

8. Dewey, J. (1938). *Experience and Education* (Reprint ed.). New York: Free Press.

9. Dougherty, D. (2012). The maker movement. *Innovations: Technology, Governance, Globalization*, **7**(3), 11–14. http://doi.org/10.1162/INOV_a_00135

10. Eisenberg, M. (2013). 3D printing for children: what to build next? *International Journal of Child-Computer Interaction*, **1**(1), 7–13.

11. Feenberg, A., and Callon, M. (2010). *Between Reason and Experience: Essays in Technology and Modernity*. Cambridge, MA: MIT Press.

12. Ferdinand, J.-P., Petschow, U., and Dickel, S. (2016). *The Decentralized and Networked Future of Value Creation: 3D Printing and its Implications for Society, Industry, and Sustainable Development.* Switzerland: Springer.

13. Gao, W., et al. (2015). The status, challenges, and future of additive manufacturing in engineering. *Computer-Aided Design*, **69**, 65–89.

14. Gershenfeld, N. (2008). *Fab: The Coming Revolution on Your Desktop–From Personal Computers to Personal Fabrication*. Basic Books.

15. Gibson, D. I., Rosen, D. D. W., and Stucker, D. B. (2010). Direct digital manufacturing. In *Additive Manufacturing Technologies.* Springer, pp. 378–399. Retrieved from http://link.springer.com/chapter/10.1007/978-1-4419-1120-9_14

16. Greenberg, J. M. (2010). *From BetaMax to Blockbuster: Video stores and the invention of movies on video.* Cambridge, MA: MIT Press.

17. Hopkinson, N., Hague, R., and Dickens, P. (2006). *Rapid Manufacturing: An Industrial Revolution for the Digital Age.* Chichester: John Wiley & Sons.

18. Ihde, D. (2008). The designer fallacy and technological imagination. In *Philosophy and Design: From Engineering to Architecture.* Netherlands: Springer, pp. 51–59.

19. Killi, S. (2013). *Designing for Additive Manufacturing: Perspectives from Product Design* (Doctoral dissertation). Oslo School of Architecture and Design, Oslo.

20. Killi, S., Kempton, W. L., and Morrison, A. (2015). Design issues and orientations in additive manufacturing. *International Journal of Rapid Manufacturing,* **5**(3–4), 289–307.

21. Koen, P., et al. (2001). Providing clarity and a common language to the "fuzzy front end". *Research-Technology Management,* **44**(2), 46–55.

22. Kolb, D. A. (1984). *Experiential Learning: Experience as the Source of Learning and Development.* Englewood Cliffs, NJ: Prentice-Hall.

23. Krugman, P. (1980). Scale economies, product differentiation, and the pattern of trade. *The American Economic Review,* **70**(5), 950–959.

24. Lindtner, S., Bardzell, S., and Bardzell, J. (2016). Reconstituting the utopian vision of making: HCI after technosolutionism. In *Proceedings of the 2016 CHI Conference on Human Factors in Computing Systems.* ACM, pp. 1390–1402.

25. Lipson, H., and Kurman, M. (2013). *Fabricated: The New World of 3D Printing.* Indianapolis, IN: John Wiley & Sons.

26. MakedEd.org. (n.d.). Maker education initiative: every child a maker. Retrieved November 23, 2016, from http://makered.org/

27. MakerBot. (2011). MakerBot thing-O-matic 3D printer kit - MakerBot industries. Retrieved November 28, 2016, from https://web.archive.org/web/20111223112406/http://store.makerbot.com/thing-o-matic-kit-mk7.html

28. Markillie, P. (2012). Manufacturing: the third industrial revolution. *The Economist,* *21.* Retrieved from http://www.economist.com/node/21553017

29. No Isolation AS. (2016). No isolation. Retrieved October 27, 2016, from http://www.noisolation.com/#home

30. Norman, D., and Verganti, R. (2012). Incremental and radical innovation: design research vs. technology and meaning change. *Design Issues,* 1–19.

31. Oxman, N. (2010). Material-based design computation (Doctoral dissertation). Massachusetts Institute of Technology.

32. Petrick, I. J., and Simpson, T. W. (2013). 3D printing disrupts manufacturing: how economies of one create new rules of competition. *Research-Technology Management*, **56**(6), 12–16.

33. Pham, D., and Gault, R. (1998). A comparison of rapid prototyping technologies. *International Journal of Machine Tools and Manufacture*, **38**(10), 1257–1287.

34. Pine, B. J. (1999). *Mass Customization: The New Frontier in Business Competition* (New ed.). Boston, MA; London: Harvard Business Review Press.

35. Ratto, M. (n.d.). Making at the end of nature. ACM Interactions. Retrieved September 20, 2016, from http://interactions.acm.org/archive/view/september-october-2016/making-at-the-end-of-nature

36. Ratto, M., and Ree, R. (2012). Materializing information: 3D printing and social change. *First Monday*, **17**(7). Retrieved from http://journals.uic.edu/ojs/index.php/fm/article/view/3968

37. Rayna, T., and Striukova, L. (2014). The impact of 3D printing technologies on business model innovation. In *Digital Enterprise Design & Management*. Heidelberg: Springer, pp. 119–132.

38. Roozenburg, N. F. M., and Eekels, J. (1995). *Product Design, Fundamentals and Methods* (New ed.). Chichester; New York: Wiley-Blackwell.

39. Sasson, A., and Johnson, J. C. (2016). The 3D printing order: variability, supercenters and supply chain reconfigurations. *International Journal of Physical Distribution & Logistics Management*, **46**(1), 82–94.

40. TCT Magazine (2013). Chuck Hull: inventor, innovator, icon; the story of how 3D printing came to be. Retrieved from https://www.youtube.com/watch?v=yQMJAg45gFE

41. Troxler, P. (2016). Fabrication laboratories (fab labs). In *The Decentralized and Networked Future of Value Creation*. Springer, pp. 109–127.

42. von Hippel, E. A. (2016). Free innovation.

43. Walter-Herrmann, J., and Büching, C. (2014). *FabLab: Of Machines, Makers and Inventors*. Bielefeld: Transcript-Verlag.

44. Wohlers, T., ed. (2009). *Wohlers Report*. Fort Collins.

45. Wohlers, T., et al. (2004). Rapid prototyping. Tooling & manufacturing: state of the industry. *Annual Worldwide Progress Report*.

46. Harris, M. (2016). Why your next car should be 3D printed: Backchannel. Retrieved February 15, 2016, from https://backchannel.com/why-your-next-car-should-be-3d-printed-63dd19ff0254

Chapter 3

AICE: An Approach to Designing for Additive Manufacturing

Steinar Killi
Oslo School of Architecture and Design, Oslo, Norway
Steinar.Killi@aho.no

Medical equipment has become somewhat the holy grail of 3D printing. It all started with customized hearing aids back in 1999–2000; today it is a multimillion-dollar industry.

In 2002 I was approached by a surgeon, Bjørn Iversen, who had worked with hip replacements for some 20 years. Over these years he had seen how the low success rate had been quite steady. A staggering 78% of all hip replacements are misaligned, resulting in discomfort and, at worst, the joints jumping out of position, requiring new surgery. As Dr. Iversen stated, "This is not fine brain surgery; a hip surgeon needs to be strong as a bull and with approximately the same intelligence." Not so sure about the latter part of the statement, but on seeing some videos of a hip surgery, something I don't recommend people with bad hips to do, it became obvious that this indeed was hard work. The rather rough way it's done also indicates

Additive Manufacturing: Design, Methods, and Processes
Edited by Steinar Killi
Copyright © 2017 Pan Stanford Publishing Pte. Ltd.
ISBN 978-981-4774-16-1 (Hardcover), 978-1-315-19658-9 (eBook)
www.panstanford.com

why the failure rate is so high; all measurements were made on site by the surgeon's more or less trained eye. Quoting Dr. Iversen, "If you are going to get your hip replaced, get that butcher-type surgeon that has done this several times, preferably successful, not the professor doctor who has done it once or twice."

What Dr. Iversen had invented was a technique to measure the position of the existing hip and then to use these measurements when operating in the new hip joint. He had some preliminary designs with him, and we started to develop a system to be used for hip replacements. At this point the device was supposed to be manufactured with injection molding, and we used rapid prototyping machines to test out different concepts. Not many weeks into the project a challenge emerged; Dr. Iversen brought in a steady stream of different joints—all would need a dedicated design. It became more and more clear that the production cost of the rather complex design would be high and due to all the different joints, the possible number of sales would rather be by the hundreds than by the thousands.

We then started to look into the possibility of actually producing the measuring devices, now named "anteversionheads," in the same material and with the same technology we made the prototypes—polyamide, which was approved by the FDA to be in contact with human tissues, and selective laser sintering, a method that due to the high temperature and closed atmosphere would be possible to use for this purpose. At this time a very successful project using rapid prototyping equipment for end user purposes had been on the market for a while; the hearing aid, in short, a silicon mold, was made from the patient's inner ear, scanned, and then produced using SLA. The electronics were mounted in the part and shipped off to the patient. This has become the benchmark and contains the three requirements for a successful 3D printing project: (1) it has to be small in size; (2) it has to be complex—complexity is for free; and finally (3) it should be expensive; producing parts that will compete with something that cost 2 cents to produce will never be an interesting market.

This moved the project on. Dr. Iversen kept coming in with different versions of joints, getting the numbers of variations up to several hundreds. Figure 3.1 shows some of the different prosthesis and two possible designs of the anteversionheads.

Figure 3.1 Seven different hip joints and two examples of anteversionheads. The fork-looking part is supposed to be attached to the anteversionheads in the grooves going in a cross, making it possible to move the prosthesis when it is mounted to the human body. Pictures: Steinar Killi and Inger Steinnes.

The decision to change from injection molding to 3D printing (not a term we used at that moment) first made the design process much simpler. We could add in supercomplex features and easily change the design for the different versions of joints. However, problems also rose. The anteversionheads were to be mounted on a polished steel shaft, and really tight tolerances were necessary for the anteversionheads to stay correctly in place, which were of some importance. Parts made by selective laser sintering (SLS), SLA, or fused deposition modeling (FDM) are struggling when it comes to tolerances under 0.1 mm. Typically at that time one would talk about +/−0.15 mm, something that was unacceptable. We sort of solved that in the design, making it flexible to accommodate the poor tolerances in the technology. This, although, together with other details of the design resulted in another, much harder to fix, problem; the parts were extremely hard to clean, or in other words, small cavities and the fact that a steel fork was mounted on the plastic part made small pieces of unsintered powder drizzle. This could of course not happen during a surgery. It is with great remorse I have to say that this story does not have a happy ending. At the time, neither the technology itself nor the design solved the issues raised above satisfactorily. The project was abandoned around 2006. This incident did, however, spur the interest in what was then labeled "rapid manufacturing," and retrospectively we could see how this project could have been successful.

There were several errors made, and the technology was not mature enough at this time, but the lessons learned that are still valid today would be:

- If a product is designed for injection molding, shifting to additive manufacturing as a production method does not mean it gets easier/cheaper. We should have restarted the design process from scratch, defining the constants (the measuring angle), and related this to the changing parameters (the different prosthesis). Instead we continued to change a design originally made for one type of prosthesis with injection-molded tolerances.
- The actual production process, with the necessary postprocessing (removing loose powder), strongly influences the design process. The different design choices should be tested at production terms, meaning how easy it is to produce a part, not only printing it, but also doing a quality check and postprocessing it, all the way to packaging it. The gap between prototyping that was successful and the production standard was not filled. The design should have addressed these issues, too.

If we look at the development of a product you can purchase on the market it consists of three stages: The first is the creative stage, often called the design stage; the actual shape and possible functions are defined. Next, this design usually goes through an engineering phase. The type of materials—fasteners, thicknesses, parting lines, etc.—are defined. This moves the design into the final stage—production. Here the tools are adjusted, possible jigs are made, and all postprocessing before packaging and delivery is solved. We could make a simple visualization of this, as shown in Fig. 3.2.

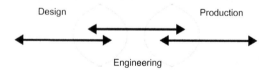

Figure 3.2 From an idea to a product, three possible phases of product development. All three phases influence each other, and knowledge about all phases is necessary; however, the processes are destined to move in one direction. Going back to the drawing table from production would be extremely expensive. Drawing: Steinar Killi and Inger Steinnes.

To what extent these three phases merge will depend on the type of products and how the product development is defined. In some companies, like automotive, everything happens within the company—the design, engineering, and finally production—although a car factory today is more of an assembly line and parts produced elsewhere arrive just in time for production. The designer will often follow the concept all the way through to production. However, he or she will rely more and more on the experts in the different phases. This will be even more so if the production is outsourced to some other place, sometimes also including engineering, mold making, and fasteners. One of the key factors of 3D printing is a lean process; the designer follows the whole process, including the last part— production and packaging. We now have a loose merge of all three phases (Fig. 3.3).

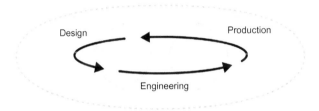

Figure 3.3 The trickle-across model. The different stages are still recognizable and have to be addressed, but it will be more holistic and adjustable along the whole process, both directions. Issues in production, for instance, postprocessing, should trickle back to the design and/or the engineering.

As seen in Fig. 3.3 actions and reactions go back and forth through the whole process, faster, and without possible problems related to cross disciplines, for example, between designers and the production people. We could label this as a trickle-across model: issues are not halted, diverged, or paused but trickle seamlessly through the process.

Over the last decade or so, almost all publications advertising with designing for additive manufacturing have focused on the midphase and some on the last phase; the shaped concept exists, but how could it be optimized through 3D printing? Examples of this are many, and you could observe it in all the conferences and 3D printing shows around the world (more about this in Section 3.1.3). On the other hand, there are products where the shape still is not developed; something has to happen in the beginning of the first phase. A third example is where a design appears through the development of a construction or an assembly. We could look at an example of this; in 2005 we were asked to design and produce an installation to be placed in front of the University of Oslo as an entrance to a public outdoor science fair. One of the criteria was to implement 3D printing in the installation. A team of architectural students and teachers developed an installation made of three triple helixes, with more than 50 different joints. The plan was to use aluminum tubes and custom-made, 3D-printed joints. Figure 3.4 shows the finished triple helix. As can be seen, the installation was made but never placed in the public area. The reason for this was we discovered some of the joints had some weak layers, and we could not risk placing it where people could be harmed, with kids climbing in it and so forth. The joints, shown in Fig. 3.4, had a shape automatically derived from the shape of the tower. Ninety-six different joints were printed for this installation. In this case, the design was the tower; the joints were engineered and then produced. The tower could of course have been produced using some other joining systems, using bolts and plates, but this was a much leaner way to do it.

Figure 3.4 A close-up of one of the 96 automatically generated joints and the structure set up in a safe environment. Design and pictures: Espen Bærheim et al.

So, how could the design process to accommodate these different ways of developing a product work? In my PhD from 2013 an approach to do this was developed—the adapt, integrate, compensate, and elongate (AICE) approach. This has been developed further and will be presented in this chapter.

3.1 AICE: An Operational Model

With the prosthesis project, described in the beginning of this chapter, and several other projects in mind, a need to develop a more holistic approach that could address not only the potential but also the challenges, visualized in Fig. 3.3, emerged. It could be adjusted to work as a model. There are several ways to visualize this; the latest version is the onion model.

The approach consists of four layers, each with individual features but deeply dependent on each other. Results are fed outward in the model, and the feedback trickles back. The first layer, *adapt*, presents suggestions to how creative, analytical, operational, and other methods could be adapted to additive manufacturing, not only as a means of production, but also as a facilitator through the whole process depicted in Fig. 3.3. This will be elaborated in the next section. The next layer, *integrate*, presents suggestions on how the engineering and production process should/could be integrated in the design process, giving instantly and constantly feedback to the developing of the shape. This section is largely a compilation of different methods and approaches developed and presented by many contributors in this field for many years. The *third* layer, *compensate*, is the moderator: how could deficiencies in process, technology, brief, and economy (the whole ecosystem) be compensated, balanced, and restructured? This section is a map that envisions the elastic boundaries of 3D printing as a product development paradigm. The final layer, *elongation*, addresses the lean possibilities in 3D printing, with redesign as a continuous process, whether it's custom made or during cocreation. This section presents models of how this could work. Let us start unpacking the onion.

3.1.1 Adapt

> *It is not the strongest of the species that survives, nor the most intelligent that survives, it is the one most adaptable to change.*
>
> **—Charles Darwin**

3.1.1.1 Design thinking

Over the last decades the term "design thinking" has evolved. Cross [1] and Verganti [2] both advocate for a "designerly" way of working.

This is a huge topic and of interest to many disciplines beyond design. Some of the key factors in design thinking could be summarized in the following bullet points:

- Have empathy for the user. The user, or more correctly users, should be in the center for all product development—not only the end users but also those in contact with the product all the way from production to sales.
- Perform tests early and keep doing them throughout the process. Feedback is crucial; waiting too long after tests could be expensive. Do iterations. "Fail fast and fail early" is a rather new proverb describing this way of doing product development.
- Use models, prototypes, and visualizations early and benefit from valuable meetings with stakeholders coming from other professions. The old proverb "A picture tells a story of more than a thousand words" goes for physical objects, too. Seeing, touching, and testing are extremely strong facilitating tools in cross-disciplinary meetings.
- Think holistic. The wider you understand the impact of a product or a situation leading to a product, the more you will gain during the development process.

When designing for 3D printing, adapting to this way of thinking could be of great benefit. Next I will show how this could be implemented.

Within industrial design numerous methods for product development and operational methods have been developed that give an overview to the whole process—what should be done first, second, and so on. There are more specific methods, for instance, to analyze users and behavior, and finally creative methods to develop new concepts. Experienced designers often use a set of methods and float between them unconsciously, and most methods could profit from new technologies, new insights into human behavior, and so forth. To visualize how these methods could be adapted to a more holistic 3D-printed paradigm we have developed a model. Figure 3.5 visualizes different methods used during product development. There are literally hundreds of design methods [3, 4], and new ones are being developed constantly. We have picked some methods, used and refined over years both through practice and in education. These again have been scrutinized for their usefulness when designing for 3D printing. The result has been mapped out in Fig. 3.5. This of course means this is not the complete overview of useful methods for 3D printing; one should look at different methods and see if it's possible to adapt them toward 3D printing. The following aims to give an insight into how this could be done through examples.

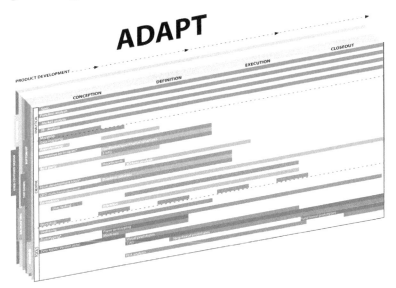

Figure 3.5 A compilation of methods used in the design process. The operational methods lie as translucent sheets on top of each other, labeled as folders on the left side. The different methods, analytical and creative, are suggested to be used at different stages for the different operational methods. Not all methods will be used; this will depend on the project type. This model could be seen as a toolbox of possible methods.

Design methods exist on many levels and in many formats, from operational methods that define a path to do the whole process, from start to end, to specific methods developed for a limited specific use. Big companies very often define their own paths, on the basis of how the companies are structured and what kinds of products they produce/design. There are, however, some more generic models developed that could serve as a base for both companies and a new way of producing, like 3D printing. These generic models lie like background folders in Fig. 3.5. These operational models will be presented thoroughly in the next section. The next level, when an operational method is selected, includes more specific methods, often developed to be used at a specific time in a design process, for instance, brainstorming. These methods occur at different times during a design project and can be seen as colored blocks in Fig. 3.5. Further, a timeline for a design process could differ immensely and there exist countless ways of depicting it. We have chosen a quite generic way of presenting it: conception, definition, execution, and closeout. These are often overlapping, and in a real process one will see loops between them.

To connect the methods to 3D printing there is an evaluation of the adaptivity to 3D printing for the models selected. This is visualized in Fig. 3.5 as starts. How and why these methods are perceived as adaptable to 3D printing is explained in the different sections. Finally, the different types of models, negotiotypes, visiotypes, etc., introduced in the next section and in Chapter 4 indicate how and when different methods could benefit from these types of models.

3.1.1.2 Multitypes

"Prototypes" is a well-known term coined long before 3D printing emerged. Basically it's a term covering everything from a simple mock-up of cardboards or even a draft on a paper if it's something immaterial to something that both looks and works like the real thing. In the design process a range of prototypes of different complexities, are made. Usually, these prototypes become more and more specific closer to the final solution. However, the actual use of prototypes differs. Jan Capjon [5], a practitioner and academic in design, investigated how prototypes could improve the design process, particularly when decisions were to be made. In short he developed four different categories of prototypes, or multitypes: *visiotypes*, *negotiotypes*, *functional prototypes*, and *seriotypes*. These names indicate a linear process, and the seriotype here is close to the final product, both in looks and in function. Multitypes could be arranged differently, and different variations of the types could be developed. I have changed this model and included *zerotypes*. As the name indicates, these are the first tangible examples of possible concepts. Zerotypes are rough and simple, maybe just a mock-up made of available parts, and there should be a lot of them. Next, in the visiotype stage, some concepts are focused and developed to a stage where not only mere function is presented. Placing of components and actual physical appearance are the key for visiotypes. The visiotypes could then be developed further to be shown in cross-disciplinary working groups. These groups often consist of decision makers, engineers, market people, etc. Here the negotiation starts: what components work well, does it fit the company brand, what will the price be, etc. Typically, these models need to facilitate these discussions, hence the term "negotiotypes. " Next, in the functional prototype stage the prototype should be possible to test. The design process has at this stage probably reached the point of no return. The test for function and minor things could alter how the final product will be. Capjon had a possible last version, seriotypes. These could be a test version made in a limited amount for user testing on a larger scale. At this point, after sufficient rounds of prototypes, in a regular product development process, the actual production process starts, sometimes at totally different places from where the development was performed. Since this book is about producing end products with 3D printing we could add one last type: *releasetype*. This type would in many cases be replacing the seriotype. What the consumer purchases is the releasetype, so this is the only type that is commercial. These terms will be explained and exemplified further both in this chapter and in Chapter 4.

3.1.1.3 Models describing a typical design process

There are several methods or models that describe how a design process could or should be executed. In short there is some kind of timeline and steps to be performed during the timeline. In Fig. 3.5 the timeline is described along the *x* axis and with some milestones, called conception, definition, execution, and closeout. This could be described with fewer words, other words, or more words, and other, more familiar ways to describe the timeline should work fine. In fact, the readers are encouraged to build their own models, changing the methods according to personal preferences. In Fig. 3.5 there are several methods that cater to the whole product development process; these are the folders on the left side in Fig. 3.5. I will present two of them and describe how and why they are adaptable with 3D printing. The readers are encouraged to try this with the other ones.

3.1.1.3.1 *Double diamond*

The double-diamond model is quite a generic visualization of how many designers intuitively work, often without being aware that such working method has a name. The British Design Council named the method as a part of its effort to develop strategic design tools. As we can see in Fig. 3.6 each of the two diamonds shows a process containing an explorative phase, investigating possibilities, and existing solutions, which at some point converge and decisions are being made. These decisions open up a new, explorative phase but this time by conceptualizing solutions. At some point a concept is chosen and finalized, this being the second converging phase. There exist many variations of this: triple diamonds, which add one more iteration, or a Christmas tree model, where the model should be read from bottom up and where the exploration and convergence phases are described less symmetrically, looking more like a Christmas tree.

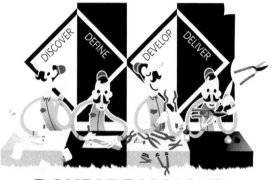

DOUBLE DIAMOND

Figure 3.6 An operational design method explaining the different phases in a design process: opening up for ideas, closing in for a definition, and opening up again for possible solutions before finally closing in for delivery. Drawing: Gjermund Bohne and Steinar Killi.

If we look into the different phases of this model, picturing the double diamond covering all phases in the timeline in Fig. 3.5 (concept, definition, execute, and closeout), different methods for different phases, described as colored blocks, could be used. Further, and this is where 3D printing comes in, along the x axis at the bottom of Fig. 3.5, we could see how the different types (zerotypes, visiotypes, negotiotypes, etc.) could be utilized at the different stages. To make functional prototypes toward the end of the process is of course obvious, and it has been done for a long time. Visualizations of different concept iterations (see Fig. 3.6) is also obviously beneficial for 3D printing. This would typically be in the overlap between visiotypes and negotiotypes. 3D printing could of course be of use even earlier in the process, when visualizing existing solutions. Working on these, using what is labeled "zerotypes," we could keep a format of visualizing ideas that becomes familiar through the developing process. When using design methods like the double diamond and others, it is often beneficial to personalize them, figuring out how the methods are useful for your process. This is not a clinical, medical process that needs to be carried out to the finest letter.

3.1.1.3.2 *Backcasting*

The different models could easily be seen as mind-sets; each of them has a timeline, and again there are steps that need to be addressed. However, we can manipulate the process, trying different angles and even altering the timeline. Figure 3.7 visualizes a method called *backcasting*. This method has typically been used for tasks that are hard to achieve, but we have a clear vision of what it is [6]. It could be described as reverse engineering: you have something that works and you open it up and see what makes it work. In other words, you start with a vision of a solution and you track the possible steps backward to where we are now. That means you will have to identify several needs that have to be addressed and solved on the path from the vision to the starting point. This would again indicate that at each step, possible solutions should be presented. When the path is made, from the vision to the present time, one should be able to trace the actual path to move. As Fig. 3.7 shows, you pick the stepping stones indicated when moving from the vision, trying to pick the best route. Very often this method will indicate one or several leaps in technology or in political decisions. This method is frequently used to derive solutions for climate challenges. The global rise in temperature has to be contained and not exceed 2° before 2050, for instance. This is a hairy goal but also clear. The next step would be to localize what causes the rise and then how this could be addressed. This method is transferable to product design. We could envision an end product, then unpack this end product and see what type of technology has to be developed, and so forth. In the car industry, companies are working with concept cars. These could be seen as indicators of a possible direction that the companies are heading, for instance, new security devices, electric cars, self-driven cars, and so forth. By presenting the vision before the actual model is made, and probably before several of the issues presented are solved, an organization creates an anticipation in the market, could check whether the concept is of interest, and could develop clear goals for itself. Usually, however, concept cars are never realized the way they are presented. The process will change it, but we will see bits and pieces of the initial visions.

BACKCASTING

Figure 3.7 Make a vision of a future goal, track the possible paths back to where you are at present, then choose the best path to realize the vision. Drawing: Gjermund Bohne and Steinar Killi.

In this case the beginning of the timeline is both the start and the end. The zerotypes would work as a visualization of a desired solution. Let's look at an example.

In 2009 a student, Andre Lyngra, developed a highly futuristic concept for a car (see Fig. 3.8). This concept is still miles away from realization, and it will most probably never be realized either. However, the scaled 3D model depicted a vision. Engineers could start discussing how hubless wheels could solve one of the issues and so forth. In this case, the multitype model is turned almost upside down; the model serves as a vision. Since it is physical/tangible it actually also serves as a negotiotype; hence engineers come up with practical solutions. This concept starts with a vision, and someone who would like to actually produce it would have to solve all the issues that rise when you start "opening" the design, tracking back to the technology we have now, and defining what technology has to be developed.

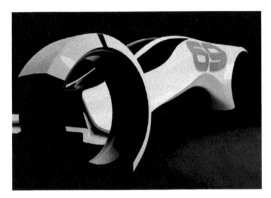

Figure 3.8   The 69 performance car. Design and picture: Andre Lyngra.

3.1.1.4   Methods used during a design process

As previously stated there exist hundreds of methods that could be of great help at different stages in a design process. When and how they are used is typically based on personal or a company's preferences. However, these methods are developed to be more specific and to address typical reoccurring challenges in the design process. Again, as most methods they are about mind-sets, to abstract an issue to increase the amount of possible solutions or to organize and systemize certain events. In the following sections, we will look at five such methods and how they could be adapted to 3D printing.

3.1.1.4.1   *User, user situation, and way of use*

Following the statements made from design thinking earlier in this chapter, the user is at the center of a designer's mind-set. We could open up the term "user" and derive a method from this. Figure 3.9 depicts the three questions one should ask: Who is going to be the user of the new design? When and/or where should it be used? And finally, how should it be used? These three questions could derive different answers, but the answers should give the project direction. These three questions are central in what we could call user-centered design [7]. By asking these questions a list of demands could be generated, which will then give input to a creative process. This method is typically what we would label as an analytical method, and although it is the method most design projects start with, it is also the method to check back with: is the concept really for kids between 7 and 10? When designing for niche markets, there is a good possibility to narrow the answers to these questions considerably, hence starting the design process with a clear direction. If we define the user group to be a limited group of people it will have implications for the following design process. It is important to emphasize again that this method is one all designers should be aware of; it is really the core of design practice.

USER   **USER SITUATION**   **WAY OF USE**

Figure 3.9   Who is the user? Where/when should something be used? How should it be used? Drawing: Gjermund Bohne and Steinar Killi.

3.1.1.4.2 *Quantitative structures and functional analysis*

A method developed by Tjalve [8] is typically in use during the early stages of a design process. The core of the method is to divide a product into volumes or structures that are necessary. Figure 3.10 shows an example of how this method is used. For instance, if you are to develop a new handheld flashlight, it will probably consist of batteries, lightbulbs, a switch, wires, a reflector, and a casing where all parts are mounted. There could be even more parts and of course a combination of parts. When defining the structures and giving the volume, it is possible to rearrange these parts in several ways on the basis of their size, functionality, and dependency toward the other parts, hence opening up for a more creative process by abstracting and deconstructing a possible design. In this case it is rather easy to see how 3D printing could be implemented; different structures at different levels of development could easily be produced, sizes and connection points could be tested, and again, since the final type, the releasetype would be derived from this production process.

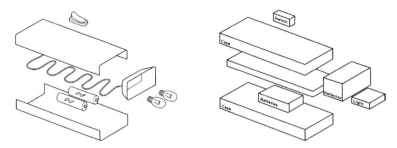

Figure 3.10   Example of what a flashlight consists of and how this could be abstracted to quantitative structures. Drawing: Inger Steinnes.

3.1.1.4.3   *Diversity abstraction forced choosing method*

The diversity abstraction forced choosing (DAFC) builds on the previous method, when deconstructing a possible part into necessary functions and/or structures one would aspire not to design what already exists, due to some default mechanisms. For instance, if we look at the flashlight again, we could combine the different parts in many ways, creating a new flashlight, but could we take it one step further? If we isolate each part and then abstract the use of that part isolated from what it's intended to do, what new shapes and functions could be developed? Figure 3.11 shows an example of this method in use. Each functional structure has been abstracted to new shapes, and then all the different possibilities have been put in a table. The next step would be to see how we could put together the flashlight using the different abstractions. The possible paths are indicated in Fig. 3.11. This method again opens up great possibilities with 3D printing, experimenting with several variations of each structure in a physical format, generating an abundance of possible new designs. This method is of value, especially when combined with the different ways of generating shapes (see Chapter 5).

Let's have a look at an example. The following case spans over several methods and was run quite systemically. Twelve master's students (industrial design) were given the following task:

Develop a drinking container with the following demands: The container should be made using a clay-based liquid deposition modeling (LDM) 3D printer and then burned and glazed. The container should address grip, way of drinking, placement, and a set volume (300 mL). There should also be a parametric input; this could be influenced by almost anything—size of user, sound, cultural aspects, etc.—different variations should be derived. Further, the container should be developed using two set methods, the above-mentioned and bottom-up codesign, presented in the next section. Most students also used the double-diamond and the user-centered method as a starting point (Fig. 3.9).

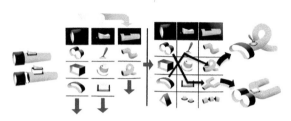

THE DAFC METHOD

Figure 3.11 The diversity abstraction forced choosing (DAFC) method, a creative version of a quality function deployment (QFD) analysis. Drawing: Gjermund Bohne and Steinar Killi.

Figures 3.10 and 3.11 show two examples of how this method could work. During this process simple prototypes, which could typically be labeled as zerotypes/visiotypes, were printed using the clay printer and also some plastic printers, Ultimakers, due to availability of the clay printer. Going from 2D visualization, that is already a great leap compared to simple text, to tangible objects printed continuously over a period of just one day brought the design process from the early, very often hard, creative phase to the define and deliver phase in a quick and fun-filled way. Figures 3.12 and 3.13 show some of the visiotypes printed in this first stage.

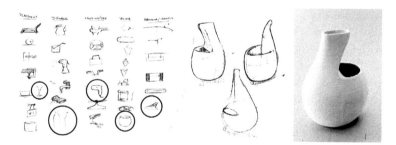

Figure 3.12 Using the DAFC method to develop a drinking container. Zerotypes and a visiotype were printed. This model went into the next stage, described in the bottom-up codesign section. Design: Jomy Joseph; pictures: Jomy Joseph and Inger Steinnes.

Figure 3.13 Another example of a drinking container conceptualized using the DAFC method. Design: Hsuan-Han Chen; pictures: Hsuan-Han Chen and Inger Steinnes.

With a concept ready as a digital file as well as physical artifacts, the next method was presented. In this case the next method was building on the first, the goal being to refine the concepts developed. However, the following method could also start as a creative starting method.

3.1.1.4.4 *Bottom-up codesign*

Codesigning, codeveloping, etc., have gained increased interest among designers and others, especially within service design. The value of bringing in more people in a design process has caught momentum, and people are core elements in several design methods, like At-One. Some companies developing physical products have also seen benefits in involving their customers in the product development process. An example here is Converse. Several models from this brand are a result of such collaboration. For Converse this is also seen as an important part of its brand building. The idea of codesigning could be manifested in several ways, and for this book it is physical products that are of interest. So, how could codesigning be beneficial when designing for 3D printing? This will be treated more in depth in the last section of the AICE: elongate. However, there is a method developed especially for developing physical products using codesign. Again, these methods could probably be found in several versions with different names. I first encountered this method in a workshop held by a Norwegian designer, Baard Arnesen. What makes this model especially valid to adapt to our purpose, 3D printing, is the early and extensive use of digital tools, or more precisely computer-aided design (CAD) tools. To get the most out of this method there are some prerequisites; there should be no less than three and no more than five participating; further, one, preferably the one with the deepest understanding of the project, should be proficient with a CAD tool. The last thing needed is a whiteboard, some markers, and a video projector. The setup is as displayed in Fig. 3.14. The one sitting at the computer is the "secretary" in the project, and that person should just translate the drawings made by the others to a digital format. The others develop concepts on the whiteboard; the secretary traces them and digitizes them, and then iterations are made; drawings on the digital file are changed accordingly. The different versions should be saved during the process. Now, with the speed of today's printers, this should be combined with prints of interesting concepts for more in-depth investigations. Changes that derive from this could then easily be carried out on the whiteboard and digitized. We are, thereby, adding an extra loop to this iterative process.

BOTTOM-UP CO-DESIGN

Figure 3.14 Bottom-up codesign: Two or three designers rework an initial design continuously, while one designer serves as the secretary, following up on the different versions in real time by projecting the CAD file on a whiteboard. Drawing: Gjermund Bohne and Steinar Killi.

So, to take the drinking container into this method, the students were divided into teams of three, and each team refined its designs: while one member sat at the computer, the other two members worked on the concept. Figures 3.15 and 3.16 show pictures from the process in these sessions. Before and during these sessions prototypes were printed. In this case one would label these as negotiotypes as they were used in the discussions. Figure 3.17 shows a typical negotiating action.

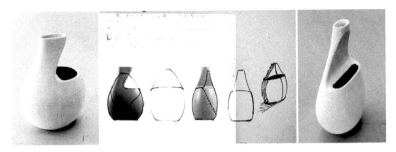

Figure 3.15 The drinking container conceptualized with the DAFC method, refined with the bottom-up codesign method. What went into this process can be seen on the left, and what came out can be seen on the right. Design: Jomy; pictures: Jomy Joseph and Inger Steinnes.

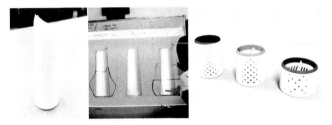

Figure 3.16 The other drinking container refined using the DAFC method; the input from DAFC is on the left, and the output from the bottom-up cocreation is on the right. Design: Hsuan-Han Chen; pictures: Hsuan-Han Chen and Inger Steinnes.

This workshop worked, of course better for some than others. Often proficiency with CAD tools became the bottleneck. A way to overcome this was making several suggestions on a picture and then letting the CAD operator make the different iterations without the pressure of performing on time. After this workshop, the concepts were going through a final refinement, and three parametric variations were derived. One of these was chosen and printed, burned, and glazed. The final result can be seen in use; it is a releasetype, although not necessarily the final product (see the end of this chapter). If we now consider the process performed and look at one of the translucent sheets presented in Fig. 3.5 and depict the actual process for the drinking containers, we can see how a typical adapted model with different methods applied could look like (see Fig. 3.18).

Figure 3.17   A typical negotiating action.

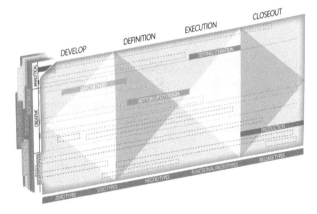

Figure 3.18   The sheet with the double-diamond method, adapted for the drinking container project with the use of three different methods. When and where the different printed types occurred are indicated in the bottom line. Drawing: Gjermund Bohne and Steinar Killi.

3.1.1.4.5   *Best party*

Again, a method to challenge your mind-set. As Fig. 3.19 shows this method is developed to break the barriers we unconsciously make early in a project. It resembles a classic brainstorming, but it is more structured. The best way to explain it is to look at its initial use: how to create the best party. Imagine you are going to arrange a party. You have a budget of, let's say, $500. You start planning, and suggestions are made, usually with an eye on the budget. Without spending too much time on this phase, you go to the next phase: how would the party be if the budget is $500,000? The fantasy is loose. Almost anything is allowed; cool suggestions hopefully fly into the discussion. When you now hopefully have the ingredients for the perfect party, at the wild cost of $500,000, you enter the next stage; given what you have found to create the coolest party, how could this be done for the original budget, $500? Now the wild, expensive, but cool ideas have to be reworked to fit the budget. The idea with this method is to open the mind-set, not kill ideas too early. If we look at this method and 3D printing, an interesting possibility arises; if one, early in the design process, uses this method and does not kill the crazy, expensive ideas too early, 3D printing is the tool that could help the wild ideas transition from stage two into stage three. This means developing new, innovative concepts really early in the process. Affordable solutions are suddenly possible with 3D printing!

BEST PARTY

Figure 3.19 Best party, a creative method to open up for innovative solutions. The aim is to arrange the best party with a tight budget. In the first phase you plan the party with a close look at your budget. In the next step you plan the party with a huge (limitless?) budget. Then, in the final stage, you decide how you could rearrange the expensive party plan with the limited budget, thereby coming up with solutions that would not even be considered in the first phase, discarded due to the cost. Drawing: Gjermund Bohne.

The intention with this section in the book has been to exemplify how methods could be adapted to 3D printing. The reader's personal favorite methods could probably also be adapted. The important issue with this part is to understand that when designing for 3D printing the notion of the possibilities that lie in this technology should be addressed early in the design process; hence *adapt* to 3D printing.

3.1.2 Integrate

Art is an attempt to integrate evil.
—Simone de Beauvoir

The next layer in the AICE model is *integrate*. While adapting the methods to generate innovative, new concepts, the next step would be to follow through. As visualized in Fig. 3.2 all three processes need to be integrated. You as a designer need to address the whole process. Of course there will be engineers and production experts necessary, but since the design is developed with 3D printing as the manufacturing method, all phases have to be catered for. Not only must mechanical and physical properties be understood and integrated in the design, the actual production and postprocessing have to be solved early. As Chapter 2 showed, all technologies have issues when being postprocessed. When looking at Fig. 3.5 again, we see that although there are steps and different methods used at different stages, we could maneuver through the different stages, integrating the tasks needed. This could be exemplified in several ways; since we are depending on digital material quite early in the process, in some formats, it would also be easy and preferable to introduce simulation tools early in the process, not just as a confirmation of a given design, but to actually generate shapes and functionality. Figure 3.21 shows how this has been used in architecture—a proposal made by Ocean North for a library in Prague [9]. The whole structure is developed using an algorithm in combination with a tool simulating the stress in the diverse beams. Several (hundreds of) iterations were generated automatically before a choice was made. So, engineering and form development were integrated but, alas, not the actual production. That would be a separate task, of course influencing the design to some extent. The library was never built though. This example could be seen as an example of procedural design, used more and more in gaming. *No Man's Sky* is a computer game that develops new worlds to visit and explore continuously, on the basis of a set of parameters, similar to the Prague library project.

ITERATIONS

Figure 3.20 Iterations could be made continuously. The drivers for the iterations could be several: human input, computer algorithms, time, etc. There could be an end solution, but the different stages could work as valuable solutions, too. Drawing: Gjermund Bohne and Steinar Killi.

Figure 3.21 Library in Prague: concept model for an invited competition. Design and photo: Ocean North.

It is quite obvious that the possibilities opening here are quite formidable, and 3D printing could definitely be one of the tools contributing to translate this way of designing to desirable products that are both unique and affordable to both develop and produce, hence solving the paradox presented in Chapter 1.

If we follow the drinking container case further, some of the projects aimed to integrate functionality into the form (see Fig. 3.22).

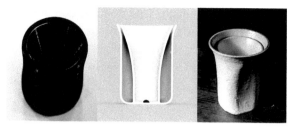

Figure 3.22 Double-walled teacup/coffee cup for better insulation. Design: Lasse Røv Thomasgaard; photos: Lasse Røv Thomasgaard and Inger Steinnes.

Some challenges also occurred when the containers went through the actual production phases: printing, burning, and glazing. Some design elements, like the legs on the container (see Fig. 3.23), were harder to integrate for production. This brings us to the third layer of AICE, *compensate*.

Figure 3.23 Teacup/coffee cup with design elements (legs) that turned out problematic in production. The leg pointing straight out in the picture on the left was built separately and mounted manually afterward, before burning. Design: Izelin Tujunen; photos: Izelin Tujunen and Inger Steinnes.

3.1.3 Compensate

There's not only emotion in the way you sing but also in what you sing. That way I can compensate it.

—Beth Gibbons

The third layer in the AICE model is *compensate*; this goes two ways. As Chapter 2 showed, the technology has deficiencies and the different technologies overlap and would need to be evaluated as such. This would imply that as the technology evolves, our choice of the technology to use would change. Even several technologies could be in use to manufacture an object. When choosing the technology to use, we would have to consider the whole process again. The need to compensate for the deficiencies will have to trickle down to the initial design process. Will the process chosen result in increased postprocessing, thicker walls, parting of the product, etc.? As stated in the previous section, *integrate*, the technology and its deficiencies have to be evaluated continuously and addressed through the whole design process. To clarify this, let's take a look at some examples.

In the beginning of this chapter there was a story about a measuring device for hip replacements that went wrong. Two deficiencies, due to the technology, were accuracy and surface quality. The accuracy was to some extent addressed, but too late in the design process. To get a better solution the design should have been totally altered, in other words, started from scratch. The other deficiency was the surface quality. The SLS process has porous products as a deficiency. This results also in a surface with loose powder. When the measuring devices were handled (mounted on a hip joint), powder was drizzling from the part. Several attempts were tested to address this, but the critical areas (those in contact with the metal, causing the plastic powder to drizzle) were semiclosed cavities (see Fig. 3.1). Today there are different ways to solve this, but one way would be to choose a technology that did not use powder, hence no loose powder drizzling. This of course would maybe lead to other deficiencies, like brittleness and not being approved as a medical material. So, similar to designing for injection molding, designing for additive manufacturing demands understanding of deficiencies and/or challenges with the technology. Compensating for these is mandatory for both. There exists an extensive amount of articles, books, white papers, etc., that go into details on this topic. In fact, compensating for deficiencies is one of the main topics if you google "designing for additive manufacturing." Conferences and technology providers churn out advisories to address this topic.

In the beginning of this section I claimed that compensate could be seen in two ways, the first being for the deficiencies of the technology. The other would be for the deficiencies in an existing process. When a company has developed a product and has started to manufacture it, there lie several small and bigger steps before the product is in the hand of the consumer. These steps could also have deficiencies that would benefit from being compensated.

The examples here are several and for a company not familiar with additive manufacturing and wondering if this could be of any benefit, this is probably where the low-hanging fruits are. A good exercise for such a company would be to evaluate any benefits from the following.

3.1.3.1   Spare parts

Of course, spare parts have been an obvious use for 3D printing. The US Army has tested moving 3D-printing vehicles in the battlefield, being able to produce spare parts on the spot, and keeping the inventory and lead time to a minimum. For companies, being able to print their own spare parts, either for their machines (not necessarily 3D printers) or for customers with heritage products could be beneficial in many ways, as for the US Army, keeping inventory and lead time to a minimum and even adding value to customers with older versions of a product. There is a legend that says Mercedes keeps two spares for every part of its older cars in stock! Whether it is still true and if so, how old the cars have to be, I don't know. However, I have benefited from it; my old convertible from 1981 was broken into and when the person tried to jumpstart it, the whole dashboard panel was destroyed. Mercedes had one in stock, the same part, even the same color, and it shipped the panel right away. The price, however, was quite steep. It is hard to imagine the inventory a 100-year-old company would need to have if it should follow this example, but the added value for me, hence for Mercedes as a brand, is substantial!

Some of the best-selling points for affordable 3D printers are the possibility to produce your own spare parts for typical household appliances made in plastic. The web is packed with more or less nifty solutions to everyday problems and also some spare parts in a format that could be easily downloaded. Whether people will actually start printing out spare parts on a large scale is probably not to be expected. However, sometimes it will be cost effective, especially if the little plastic part is crucial for a larger, expensive construction to work or the broken item has some kind of sentimental value. Some jeans/shoes, etc., we tend to repair over and over, some not. If we look back at Fig. 1.5 in Chapter 1, there is a potential here along the line between the private, low-volume prosumer and the big corporate producing for a large global market.

3.1.3.2 Production aids

If you follow a product the whole way from when the raw material enters the facility to the point where it is shipped off, there are many steps. Analyzing every step for whether it would benefit from use of 3D printing could render some interesting results! Are you using jigs and fixtures? Who makes them, and what is the lead time and cost? Could customized tools ease repairs, inspection, or other manual labor? Whether these products are being produced in-house or purchased there could be money to be saved. Fig. 3.24 shows a simple jig/*sjablong* for painting.

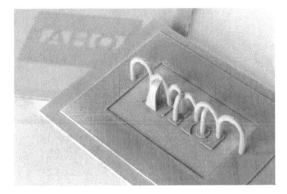

Figure 3.24   A simple 3D-printed jig for painting. Design and photo: William Kempton.

3.1.3.3 Enhancing the design

Sometimes when a company contacts me regarding 3D printing, it has a part and asks whether it would be cheaper to 3D-print it than the way the company does it today. This usually results in a negative answer. The tools are then already made, and the large investment is made. The actual production cost would almost always be lower than for 3D printing (see Chapter 1, Fig. 1.8). There are of course exceptions to this. If a complex part is milled out from a block of metal in several steps, the printing cost could be competitive. However, both to the first and the last situation there is a follow-up question: would it be beneficial for enhancing the design, compensating for deficiencies both in production and in use? A classic example of this is to investigate and enhance the topology of a product. Figure 3.25

shows the generative development of a door hinge, from a hinge made the old-fashioned way—cutting and folding plates, a functional part, far from being optimized for production (several manually working steps) or from use (probably heavier and bulkier than necessary). Another way would be via several steps using a simulation tool (here Abbacus is used) to trim the geometry, removing unnecessary material. The final design is optimized for use, much lighter, using less material, and even more cost effective to 3D-print than the original.

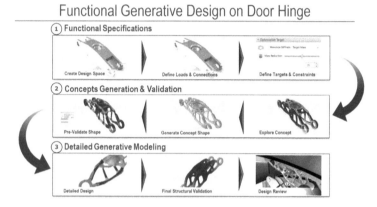

Figure 3.25 Design optimization of a door hinge. Designed in CATIA using Abbacus and Tosca for topology optimization. Design and photo: Dassault Systèmes.

Some would argue this is a new design, and in many ways it is. However, the functionality is the same, and it is not necessary to go the whole way back. Again, this could be a low-hanging fruit for a company investigating the benefits of 3D printing.

The third layer of the AICE model is *compensate*. With this we mean you have to compensate for the technological impact. When a production method becomes so dominant, for better and worse, it is a danger that it will steer the process, deriving solutions that are not optimal. This will be further elaborated in Chapter 4. For instance, in the Durr watch case, the 3D-printed releasetype was sellable, and the wise thing was to compensate for the deficiency in the material and process, changing vital 3D-printed parts with more standard-produced computer numerical controlled (CNC)-milled aluminum. However, the compensate layer goes further. During production preparation, what could also be labeled the engineering stage, there are possibilities to compensate for deficiencies in classic production methods, like injection molding. This area is somewhat better researched than the previous parts, the developing of the concepts. Examples of how this could be used are many; there are brackets and hinges optimized by removing material, used as helping structures during production, for instance, jigs or fixtures. See Fig. 3.26 of a fixture 3D-printed for a milling operation. For a company doing ordinary mass production there is potential for utilizing 3D printing all through the production process, compensating for long lead time on production equipment, expensive spare parts, etc. This is of course being done already by many companies, but this is still a dormant potential.

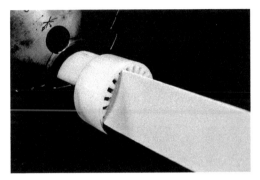

Figure 3.26 Fixture for milling. Design: Geir Jarle Jensen; picture: Inger Steinnes.

Finally, if we look at the drinking containers and the challenges that occurred, there was some need for compensating these. First, the clay printer has its obvious limitations: the overhang can't be more than approximately 30%, the support structure could be difficult to separate from the desired structure when removing, and the clay stays soft and will collapse slightly due to the weight itself. Most issues could be solved, although it would require some manual labor in the production process. One way of automating this would be to build support scaffolds with a plastic printer and use this as jigs during production. These issues are typical since we use a clay printer. Other technologies will generate other problems/ challenges. Interestingly, compared to ordinary clay making, the 3D-printed products became somewhat more naturally porous due to the way of production and this made the burnout process easier. A massive amount of clay with voids would lead to an explosion if air trapped can't get out during warm-up.

It is not only the production that is compensated, but if we look at the teacups designed by William Kempton (Fig. 3.27), the actual shape and grip are made by changing the tessellation of the digital file. The production file for 3D printing, .stl, split the model into triangles, so a sphere will actually always be faceted. If the facets are really small we perceive it as a smooth sphere. In Fig. 3.27 the settings of the resolution for the file (tessellation, smaller or bigger triangles) are altered, resulting in unique designs, although all with the same main shape and size. This could, of course, also be altered, adding more complexity to both the design and the process. The printed cups can be seen in Chapter 5, Fig. 5.5. This brings us to the last stage of the AICE model, *elongate*.

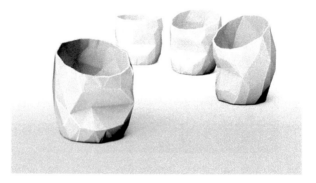

Figure 3.27   Teacups. Design: William Kempton.

3.1.4   Elongate

Pluto's orbit is so elongated that it crosses the orbit of another planet. Now that's . . . you've got no business doing that if you want to call yourself a planet. Come on, now! There's something especially transgressive about that.

—Neil deGrasse Tyson

The final layer of the AICE model is *elongate*. This way of thinking should encompass the whole design strategy. As stated in Chapter 1, following the findings in the production cost curve, there is a paradox at loose; if the number of products is too high, conventional production methods should be used; if the number is too low, the development cost divided among very few parts will drive the price through the roof. There are some solutions to this kind of paradox:

- You need to localize the sweet spot for your product. The amount produced has to fit the interval between too few to be economically viable and too many for the production method to be viable. This interval will be totally product/market dependent and could of course open up for new products that suddenly become viable economically although the price is still quite high.

- Create custom products. There will never be an interval since every product is unique, although generation of the different versions will greatly influence the price, hence making it at some point not commercially viable. If, however, the different versions could be derived more or less automatically or by the customers, there would probably not exist a maximum number of the initial designs, leading to what could be called mass customization. More about this in Chapter 6.

- The third possible solution is an ongoing, iterative commercial process. The model in Fig. 3.28 indicates how this could work. The initial process could follow the methods presented in the first part of this chapter (adapt-to methods, integrate, etc.), but when you get to the stage where you have a product that could be launched, commercially you enter a looping process. The final product does not necessarily ever have to be finalized! Input from users goes into the design process over and over, elongating the process. There are several ways this could be done, with access to private printers, print shops, etc., early, but useful and commercially interesting designs could be shared on open or closed platforms in a true codesign process. There are of course several challenges here, economy being an important one and legal issues and liability others.

There exist 3D-printed products on the market today that fall within these three categories. Hearing aids have been mentioned, and in Chapter 6 two case studies of 3D-printed glasses or eyewear are presented. In the latter case there lies a potential in testing out several different models on the market at a relatively low cost and maintaining a good margin, hence a variation of the third solution.

If we look into the second bullet point, custom products, elongate will have a quite specific meaning. When a shape is designed to change for different versions but should still be recognized as an explicit design there are guidelines that should be followed. For example, consider a sneaker. A company like Adidas does not design each size for every shoe it makes. One shoe is designed. For men this size is 10; in production this design could be stretched to size 13 or to size 7. If the shoe is going to be bigger or smaller, the design will start to be less recognizable and the designer needs to redesign especially for these sizes, increasing the cost. The key here, however, is the possibility to maintain the recognizable design from size 7 through size 13. The initial design has, in other words, an elasticity in the shape [10]! Simple forms like circles, cubes, pyramids, etc., could easily be scaled linearly up and down and even nonlinearly; a sphere could be slightly compressed and still be considered a sphere, but at some point we will struggle with that perception. Developing shapes that could be stretched and still be perceived as being the same shape could be of great value. In Chapter 6 this will be elaborated further. Hence, branding and design elements are of special interest in some of the 3D-printed products we see on the market today and in the pipeline.

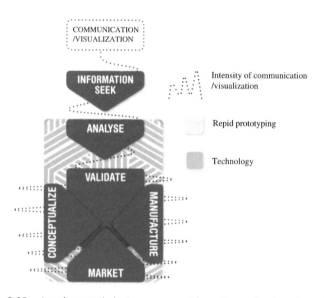

Figure 3.28 An elongated design process. There is no final end product; feedback from the market continuously influences the concept, leading to new versions or products. The four elements, conceptualize, validate, manufacture, and market, work in a loop, not as a linear process. Customization could be an example of this if it's not just fixed changes, like a size being rescaled. Drawing: Emilie Olsen and Steinar Killi.

If we again take a look at the drinking container, this issue was stressed in the design brief—a parametric input. By introducing this, the end product was from the beginning designed to change. How and why it should change varied greatly, but it opened the way for a product that could be altered on an industrious scale. Ceramic drinking containers have for ages been a popular artisan product, with unique designs and handcrafted values. One of the goals for this task, the drinking container, was to investigate if these artisan values could be transferred to an industrious production line, given the opportunities 3D printing has and with a design process that is aware of both possibilities and challenges in this process. Again, the teacup with different tessellations (Fig. 3.27) is an example of this. Printed versions of the cups can be seen in Chatper 5.

3.2 Using the AICE Model and How the Drinking Container Came Out

The different layers of the AICE model indicate the possibility to focus on one specific layer, for instance, compensate, and as with all models and methods, there is always a great potential in customizing and tweaking it to become better suited for specific needs.

All this said, the AICE model is designed to be used, or at least evaluated as a whole. The different layers will influence the others. Further, a design process should be designed, meaning this model should be adapted, tweaked, and challenged—even developed further. Product development processes will be altered rapidly in the coming years, not just because of 3D printing and other technological leaps. How we consume products will probably be just as important as how we produce them; will there be more charing, more modular thinking, less waste, less transport, more customization, and more niche products? The last chapter in this book will delve into some of the challenges and possibilities that emerges from 3D printing.

Finally, how did the drinking containers come out? This kind of stunt project went over a period of one week. The two methods were mandatory to use, and both processes and the end results were monitored toward the AICE model, as the different sections in this chapter have shown.

Not all 12 projects were successful, but in Fig. 3.29 some of the containers are in use, and have been for a long period now. Some projects will be reworked, and some will just fade away.

Figure 3.29 Some of the containers in use.

References

1. Cross, N. (2006). *Designerly Ways of Knowing*. London: Springer Science+Business Media.

2. Verganti, R. (1999). *Design-Driven Innovation: Changing the Rules of Competition by Radically innovating What Things Mean*. Boston, MA: Harvard Business Press.

3. Hanington, B., and Martin, B. (2012). *Universal Methods of Design: 100 Ways to Research Complex Problems, Develop Innovative Ideas, and Design Effective Solutions*. Beverly, MA: Rockport.

4. Kumar, V. (2012). *101 Design Methods*. Wiley.

5. Capjon, J. (2004). Trial and error based innovation (PhD thesis). Oslo School of Architecture and Design.

6. Ebert, J. E. J., Gilbert, D. T., and Wilson, T. (2009). Forecasting and backcasting: predicting the impact of events in the future. *Journal of Consumer Research*, **36**, 13.

7. Vavik, T., and Øritsland, T. A. (1999). *Menneskelige aspekter i design*. Trondheim: Tapir trykk.

8. Tjalve, E. (1987). *Systematic Design of Industrial Products*. Copenhagen: Technical University of Denmark.

9. Killi, S. (2010). Form follows algorithm: computer derived design for rapid manufacturing. *Innovative Developments in Design and Manufacturing: Advanced Research in Virtual and Rapid Prototyping. Proceedings of the 4th International Conference on Advanced Research and Rapid Prototyping* (October 6–10, 2009), Leiria, Portugal, 575–590.

10. Killi, S. (2007). Custom Design: More Than Custom to Fit! London: Taylor & Francis.

Chapter 4

The Impact of Making: Investigating the Role of the 3D Printer in Design Prototyping

William Lavatelli Kempton

Oslo School of Architecture and Design, Oslo, Norway

william.kempton@aho.no

4.1 Introduction

4.1.1 Prototyping as Design Development

A common feature in almost any design endeavor is its involvement in the making of various forms of models and prototypes. These prototypes are often described as the articulations, or manifestations, of ideas and concepts. The making of prototypes and models is an integrated part of many different creative disciplines, both in architecture and in design. However, the ways in which these prototypes are made, and with which tools, differ considerably. An architectural prototype is described by Runberger [42] as "an object of continuous investigation," both malleable and performative, that

Additive Manufacturing: Design, Methods, and Processes

Edited by Steinar Killi

Copyright © 2017 Pan Stanford Publishing Pte. Ltd.

ISBN 978-981-4774-16-1 (Hardcover), 978-1-315-19658-9 (eBook)

www.panstanford.com

allows the authors to cocreate and iterate on a physical design. In designing digital, screen-based experiences, the need for prototyping, as explained by Houde and Hill [2], is a crucial and complex task. It may involve the creation of both interactive slide shows and physical foam-core models. In product design, the importance of physical prototyping is particularity evident in the fact that end results are often physical.

While the discussion of attributes, skill, and tools in prototyping remains relevant, the question of what prototypes and artifacts actually do has been of continued interest among design researchers. Houde and Hill [22] proposed a model that tracked prototypes within three dimensions: the artifact's role in a user's life, the look and feel of the artifact, and the implementation of it. Buchenau and Suri [7] furthered the argument of prototypes as experiential components that exist in order for a design to be understood, explored, and communicated. Further unpacking the notion of prototypes as an integrated part of product development, Lim et al. [32] elaborated a theoretical framework of prototyping "as a vessel for traversing a design space" or as "purposefully formed manifestations of design ideas." Their description of design activity, which can be seen as reflective rather than prescriptive, situates the prototype as a filter and qualitative manifestation of a particular design space (Fig. 4.1).

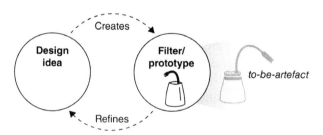

Figure 4.1 Prototypes act as filters or manifestations of a design idea. Illustration: William Kempton.

4.1.2 Making as a Critical Practice

While many of these studies emphasize the roles artifacts play as representations and filters of design intentions, other scholarly contributions emphasize making as an act for critical inquiry. Matt

Ratto (2011) coined the term "critical making" as a mode of inquiry that extends theory into physical space [38]. Ratto's theoretical framework for making emphasizes critique and exploration in order to understand new concepts, rather than solving them with technically sophisticated prototypes. Critical making can be seen as a juxtaposition of critical thinking, which is based on linguistic and theoretical expression, with tacit acts of making artifacts. In recent studies, Ratto and Ree [39] investigated 3D printing as a fluid technological phenomenon that has great implications for materializing digital and physical convergences. Critical making being a potentially socially transformative technology, Ratto and Ree employ critical-making workshops with layperson participants. This allows these design researchers to delve into topics such as literacy (the need for development of skills in a new digital economy), infrastructure (citizen involvement through making), and legislation (the potentials and ramifications of collaborative, "open-source" sharing of designs) [39].

4.1.3 Outline

In this chapter I investigate the emerging use of digital fabrication from the perspective of product design practice. 3D printing, in particular, has been primarily seen as a prototyping tool, leading to the name "rapid prototyping" (RP). As a collection of technologies for reproducing and generating physical artifacts, based on a digital blueprint [21], the emerging trend in 3D printing is a gradually movement toward production of usable artifacts, due to the gradual improvement of material properties, affordability, and speed [13]. To understand the potentiality of 3D printing as a tool for prototyping ideas and concepts, while at the same time as a platform for additive manufacturing (AM), provides the central problematic in my research. I therefore ask, "How is the emerging, expert use of digital fabrication providing new models for product conceptualization?"

This article ties into theories and discussions from product design that relate to emerging craft and practice. Whether it's labeled a prototype, model, or a mock-up, the act of making representations is an inherently important process for understanding and solving complex issues. Arguably, these acts are something designers have always been doing. What this article attempts to analyze is the developing use of 3D printing in the systematic design process. This rationale for investigating the 3D printer's role in design prototyping is to understand how it might complement the evolutionary ways in which design is being developed.

My chapter is separated into six main sections. The introduction outlines various motivations for making, with artifacts and representations, in a product-oriented design process. This is then tied onto emerging knowledge in the use of 3D printing as a developmental and production tool. Following a further discussion on the role of representations in design prototyping, I discuss in Section 4.5 how expert designers work with 3D printing, seen from the case of the development of the SunBell lamp. I then present an existing model of product design prototyping, which attempts to organize the role of design representations within design development processes. This model is then further elaborated in Section 4.6 of the article to accommodate the developing role of 3D printing as an integrated tool for development and material production.

4.1.4 Methods

As mentioned earlier, the making of artifacts and representations is often integral in the developmental process of many design practitioners. While artifacts can be described as vessels, filters, or manifestations of a design space [32], they are often complemented by methods and tools for observation and analysis. To understand latent user needs, contexts, and scenarios, designers are increasingly involved in mapping and research activities, such as gigamapping [46] and ethnographic field research [47]. As a result, the developmental process of design practice can be seen as "oscillating between poles of Real versus Abstract and Understanding versus Making" [28].

In contrast, the methods I have used to understand the contemporary phenomenon of how digital fabrication tools play a role for product conceptualization borrows from social science research methods such as case studies [51] and qualitative interviews [29]. These methods of inquiry not only serve to explore the phenomenon in question but also allow me to understand and provide qualitative descriptions of their "designerly" approaches [10].

As a designer and researcher, I also take on an observational, ethnographic role, which also has its roots in the social sciences [37, 43]. This can be likened to practices such as field research, which seek to understand practices of everyday life through a firsthand view.

As a way of addressing the phenomena of design and architecture practices as a continuous dialog between themselves and the situations in which they find themselves, Donald Schön (1983) introduced the notions of reflection-in-action and reflection-on-action [44]. Briefly put, these notions describe the way in which practitioners of design are found to reorganize their steps within a particular making process as they gather new knowledge of a situation. The designer reflects in action on the situation "talking back" to him or her, which leads the designer to form yet new strategies and moves [44].

Although the focus of my attention lies in observing and reflecting upon the developmental practices of other designers, my observational, ethnographic approach is influenced by my focus as a researcher and as a designer. Extending Schön's notion of reflection-on-action, which emphasizes the revisiting of performed actions, I combine the knowledge that is obtained by me as well as the subjects of my observation to gain insight into the role of the 3D printer in design prototyping.

This chapter forms a part of my ongoing article-based PhD design research project [41], which centers on emerging design practices in digital fabrication. Because of the wide-ranging involvement in digital fabrication by many different social groups, the role of this book chapter is to investigate the contemporary practices of expert designers.

4.2 Background

4.2.1 From Rapid Prototyping to Additive Manufacturing

The emerging roles and potentialities of 3D printing are increasingly discussed in the popular media, spanning from an economical viewpoint [49] to perspectives on sustainability [31]. Because of the increased availability of 3D printing technology both for personal use as well as through services, digital fabrication tools and technologies can be seen as a distributer of personalized material production and even carriers of cultural expression.

Throughout the 1970s and the 1980s, the advancement of computing power and development of computer-aided design (CAD) applications, which made it possible to draft and draw on a screen, made it possible for a wide range of industries to engage in a new, digital material domain. Although initially used by automotive, naval, and aerospace industries, the use of CAD has had a profound impact on design development, particularly in integrating, modeling, and evaluating design ideas [50].

The earliest technologies for fabricating digital objects with 3D printing came with the invention of stereolithography by Charles Hull [23]. This gave way to the first era of 3D printing, which centered on making prototypes and models for product development in professional industry. "Rapid prototyping," as the technologies were initially labeled, was revolutionary in that it made it possible to make multiple, precise replications of digital models that had been designed in CAD applications. Whereas the traditional way of making prototypes had been through hand work by skilled craftsmen, RP allowed product developers to make accurate prototypes in a rapid fashion [35].

The role of 3D printing in a product development process can be divided into three separate categories: the making of artifacts as representations for a concept, idea, or function; for making as a way of making tools for facilitating the making process; and lastly, as a means of manufacturing as artifacts. The last is of particular interest as digital fabrication technologies are increasing in quality, offering, and speed, making it possible to make usable, functional artifacts. The term "additive manufacturing," which sees the technology as a platform for manufacturing, is of interest for engineering-centered research [17], as well as in product design. It also suits designers who look at new means of manufacturing, as well as potential users, as a range of new possibilities for materialization are opening up.

Some of the potentialities of AM include mass customization [36], the tailoring of mass-produced artifacts to each person's preference, and the concept of enhanced design [24], where AM acts as an alternative for conventional manufacturing methods (such as injection molding) by improving the consolidation of parts, minimizing labor, and providing improved functionality.

4.2.2   Ubiquity and Stratification of 3D Printing

There is a clear demarcation between a broader and growing social interest in 3D printing and the expiry in 2009 of major 3D printing patents [12]. Of note is the RepRap movement, initiated by Adrian Bowyer, which seeks to develop open-source 3D printers for hobbyists. The movement is motivated by the idea that "the self-copying rapid-prototyping machine will allow people to manufacture for themselves many of the things they want, including the machine that does the manufacturing" [6]. Although the open-source RepRap movement has inspired much of the initial interest in developing consumer-grade 3D printers, it is often overshadowed by a plethora of commercial fabrication devices [4], online RP services such as shapeways.com, and user-friendly design programs [18]. On the basis of these continuously improving technological offerings that relate to digital fabrication, it is tempting to think that 3D printing will become a ubiquitous tool. This is also seen in comparison to the development of personal computing devices that have steadily moved from being stationary objects to laptops, tablets, and handheld smartphone devices.

In discussions on ubiquity and 3D printing for children, Eisenberg [14] discusses the importance of considering personal, digital fabrication as something more than a desktop object, much like the computational devices we carry in our pockets. Central to his critique of incorporating digital fabrication into an educational setting is to offer children "an experience of self-directed construction" [14]. This is seen in contrast to his view of children's experience of the web—with easy access to entertainment, online shopping, and even lectures—as an increasingly consumptive space.

From a multidisciplinary standpoint, Robert Ree [40] sees the idea of ubiquitous 3D printing as a misconception. He argues that because 3D printing cuts across several disciplinary boundaries, its purpose is stratified across different uses. Ree exemplifies this through a series of contexts and from an academic point of view, from which 3D printing can be utilized to teach people to open the "black box" of technologies, to that of a social galvanizer for collaborating on material investigation [40].

4.2.3 Contexts for Additive Making

The arguments advanced by Ree and Eisenberg point to a need to understand 3D printing as an emerging technological tool within a specific environment. As I am to understand how expert practices in product design are making use of digital fabrication, I next continue my discussion by referring to explorative use of 3D printing for materializing digital information by the design company Skrekkøgle.

Figure 4.2 A rich design space where tools, materials, technology, and methods meet. Favorable combinations of straps, plastic enclosures, and even sensory feedback are found through the method of trial and error, material exploration, and experiencing a messy but necessary process. At the Skrekkøgle offices in Oslo, Norway. Photo: William Kempton.

Leather straps, milled aluminum, electrical wiring, sensors, doodles, and 3D-printed watch strap buckles are spread around a large table (Fig. 4.2). Across the table, at the other end of the room, a silicone mold is seen revealing the negative shape of a figurine, while snippets of code are being transferred to a blinking Arduino board. Almost every square centimeter of available space in the room seems the victim of some kind of making experiment. The space, situated in a back alley in Oslo's Grünerløkka district, is the home of Skrekkøgle, a multidisciplinary design office run by Theo Tveterås and Lars

Vedeler. They have come to represent a generation of designers who embrace digital and physical materials through their emerging use of technology, as well as the things that are literally coming out of their 3D printers.

The various artifacts on the table are the remains of what became Durr Alpha, a sensorial wristwatch the duo created in 2014. Durr refers to the shivering sensation the watch gives every five minutes as a way of indicating time. Alpha, referring to the maturing process of software development, in many ways encapsulates the state of the product, as it was made in a limited, completely hand-assembled edition. Using 3D printing services and simple prototyping electronics, Skrekkøgle was able to release the initial product to a public audience. "We learned so much about quality control and logistics from the project. From color-dyeing the 3D printed plastics to configuring the code. As soon as we released the initial product we decided to make another, better iteration of it," Lars says as he fondles an early prototype. The following year, the Durr Beta (Fig. 4.3) was released, this time incorporating a larger extent of external assembly, computer numerical control (CNC)-milled casings and custom printed circuit boards (PCBs) fitted onto a 3D-printed board. In many ways, the naming of the watch, Durr Alpha and Beta, symbolizes a convergence between the fluid development of software and the emergence of digital fabrication technologies, such as 3D printing.

Figure 4.3   The Durr Beta watch, featuring an aluminum case with SLS-printed internals and a leather strap. Photo: Skrekkøgle.

As the boundaries between digital and physical space draw closer together, both physically and metaphorically, for designers there open up both possibilities and implications for designing. One of the possible convergences of digital and physical materials, and extended by network connectivity, is hybrid artifacts, popularly labeled "Internet of Things" (IoT).

4.2.4   Hybrid Artifacts

The computational objects and artifacts that we interact with in our everyday lives—telephones, music devices, cameras, smart watches, and quantified-self devices—can be seen as blurring the intersection of analog materiality and digital information. Such devices are continuously enriched and upgraded through new technological offerings and features. However, their presence within larger communication networks and their continuous exchange and computation of data distort what Knutsen [26] calls the "spatial context" of static artifacts. Such IoT product hybrids can be seen as breaking the mold of traditional tools and interfaces. Rather than being confined to certain domains and purposes, hybrid devices can be seen as "complex articulations and assemblages of material and cultural domain" [26].

I align my notion of a hybrid artifact between product design and AM with Knutsen's description of it as a complex assemblage of material and culture. Indeed, the material output of a digital fabrication process converges heavily between digital and physical space. As I later argue, an emerging kind of artifact— the releasetype—can be described as a hybrid artifact. It coexists as digital and physical material and comprises instances of an end product realized through AM. These artifacts are hybrid products in that they rely on interdisciplinary design competencies and processes in order to create the necessary design framework, which is adaptable by a digital interface. Such interfaces can already be seen in web-based self-services such as the NikeiD online custom shoe design service (Nike: nikeid.nike.com), which encourages creativity and allows customers to partake in codeveloping products [16].

4.2.5   Making Representations as a Way of Designing

Designers are concerned with making as a means of articulating and exploring their ideas and concepts. As with almost every design practice, such as publishing; jewelry design; and industrial-, interaction-, and service design, working with representations, such as sketches, models, and prototypes, is vital to both developing and presenting concepts methodologically. These concepts may be as diverse in both complexity and form as templates for online magazines, a mood board for a collection of necklaces, or complex flowcharts and guidelines for a new pension reform. In industrial design, a concept is usually presented as a tangible artifact so that a 3D-printed scale model or a series of prototypes represents the tactile, compositional, functional, and aesthetical qualities of a design.

In craft production, the act of designing is tightly knit with the physical making of that object [45]. A pottery maker who is turning a piece of clay is gradually making up a form as he or she goes through a process. However, in industrial and contemporary design practice, the act of designing an object is separated from the making of it [11] as the object is designed for someone else. The core goal of the industrial designer is, therefore, to provide a rich description of a design before it is produced. This entails how the product is used, how it looks, and how its manufactured. When designing a children's chair or a desk lamp, the industrial designer is not only concerned with how comfortable the chair is to sit in or if the lamp falls in line with the taste of the perceived users of it. For a design to be successful, the product should be manufactured in a rational way; it has to comply with standards of the surrounding environment, as well as hopefully fill a perceived market need.

4.3   Prototypes and Design Representations

The making of representations is arguably a vital part of the process of developing products. They encompass everything from initial sketches to elaborate drawings, models, and prototypes. In addition to facilitating the evolution of a design brief into a design solution, representations such as models and sketches act as modes of communication between the involved actors of a design process—between the industrial designer and the engineer, between the designer and the manufacturer, or even between users and the designer. In a broader view, mock-ups or cultural probes may even be used to critically engage with users, as opposed to gaining concrete solutions to user needs [5].

A holistic perspective of visual design representations (VDRs) is offered by Pei et al. [34] as a means of enhancing the communication between the involved actors, particularly emphasizing industrial designers and engineers. The authors point to the inherent cultural differences between thinking styles and values of these two disciplines as leading to misinterpretation. For instance, an industrial designer might focus on the aesthetical attributes of a sketch, while an engineering designer may use sketching as a means of solving technical details. Similarly, when making models an industrial designer might consider usability aspects, while an engineering designer might make models to evaluate mechanical principles and production feasibility.

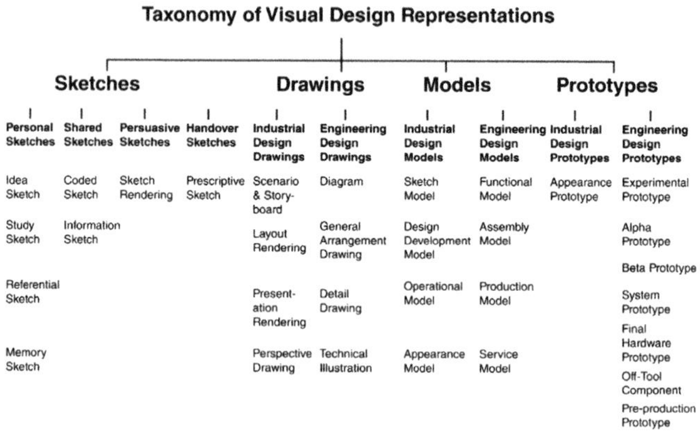

Taxonomy of Visual Design Representations

Sketches				Drawings		Models			Prototypes
Personal Sketches	Shared Sketches	Persuasive Sketches	Handover Sketches	Industrial Design Drawings	Engineering Design Drawings	Industrial Design Models	Engineering Design Models	Industrial Design Prototypes	Engineering Design Prototypes
Idea Sketch	Coded Sketch	Sketch Rendering	Prescriptive Sketch	Scenario & Story-board	Diagram	Sketch Model	Functional Model	Appearance Prototype	Experimental Prototype
Study Sketch	Information Sketch			Layout Rendering	General Arrangement Drawing	Design Development Model	Assembly Model		Alpha Prototype
Referential Sketch				Present-ation Rendering	Detail Drawing	Operational Model	Production Model		Beta Prototype
									System Prototype
Memory Sketch				Perspective Drawing	Technical Illustration	Appearance Model	Service Model		Final Hardware Prototype
									Off-Tool Component
									Pre-production Prototype

Figure 4.4 Taxonomy of design representations (Pei et al., 2011, p. 69).

The resulting taxonomy can be seen as a generic timeline, in terms of both complexity (a sketch is usually less time consuming than making a physical model) as well as evaluative purposes—an appearance model is probably easier to comprehend than a drawing of it. The VDRs (Fig. 4.4) are divided into four main types: sketches, drawings, models, and prototypes. The chart is heavily influenced by the opposing cultures of industrial designers and engineers, as is seen by the continuous comparison of the two in each type. Although it doesn't specifically refer to any particular tools for making, such as the use of hand tools or digital fabrication, it provides a clearer image of the landscape of representations in relation to the opposing disciplines.

4.4 The Changing Character of Design

Design practice varies across the various disciplines it encompasses. From engineering-driven design to fashion design, the processes that facilitate designed artifacts and outcomes intersect on several levels [30]. They often share a sensitivity toward relating and addressing users' needs in different ways, often referred to as user-centered design [1]. To address latent needs and provide satisfactory user experience, the process of design involves techniques for collecting data, observation, usability testing, and prototyping. The process of making products is necessarily a multidisciplinary one, and it is often formalized as organizational strategies.

Marketing departments, designers, investors, material experts, engineers, salespeople, and potential users are all actors in a complex web that makes up the process of product development and innovation. All these actors may perceive the act of developing products in different ways. This makes it necessary to formalize both the methods and strategies that comprise the process. Large corporations and design companies, such as Unilever and IDEO, are continuously refining these methods, strategies, and approaches in an attempt to be innovative and successful.

4.4.1 New Product Development

The overall managerial strategy of product development and innovation known as new product development (NPD) is concerned with transforming market needs into new products while emphasizing speed, flexibility, and on-time delivery [48]. The NPD strategy is generally described through a set of stages and gates [9], which typically involves discovery, idea scoping, building a business case, development, testing, and validation, before being launched. The various stages of the process call for multidisciplinary approaches, including that of marketing, engineering, and design [15].

The role of design is emphasized as conceptualizing and clarifying fuzzy or ill-defined problems (wicked problems), centered around the initial stages of the formalized NPD process, often described as the fuzzy front end [27] of design. Often seen as preceding the formal product development process [3], the fuzzy front end is handled by designers as they are skilled in transforming fuzzy problems into ideas and concepts.

As mentioned earlier, the user-centered approach requires designing practitioners to move among modes of looking, learning, asking, making, testing, evaluating, selecting, and communicating [33]. Because users often prove to have different needs and opinions, these modes of design development are often revisited in a cyclic fashion. Tools for facilitating parts of these processes, such as interviewing and observing users, are complemented by acts of making and testing, in order to further the developmental, iterative process. As technologies for 3D printing are becoming increasingly quicker, cheaper, and easier to use, so has their use in the process of making products.

The literature concerning NPD strategy often emphasizes the role of design as being in the initial, fuzzy stages of the developmental process. However, as the cases in the following sections elaborate, designers are increasingly enabled by digital fabrication to engage in product development that travels beyond the fuzzy front end.

Figure 4.5 illustrates how the modes of design practice—looking, testing, and communicating—can be viewed in relation to the multidisciplinary view of NPD. It suggests than rather than being confined to the initial fuzzy front end of NPD, design practices are in fact integrated into the entire developmental process of making new products.

How then are designers actually using digital fabrication to facilitate their work? Through two cases—the SunBell portable lamp and the DF1 ski pole grip—I will attempt to unpack the way 3D printing tools are involved in the design process, as well as provide a contextual frame of what role physical design representations take.

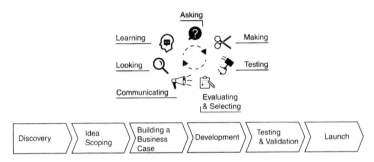

DESIGN PROCESS IN NEW PRODUCT DEVELOPMENT

Figure 4.5 The design process within new product development is iterative and affects more than just the initial fuzzy parts of the process. Interpreted from Milton and Rodgers (2013) and Cooper (2011).

4.5 Situating AM Prototypes within Design Practice

4.5.1 Developmental Prototypes

I now turn to the case of the SunBell lamp, an award-winning portable solar-powered lamp with a USB phone charger. It is designed by the Norwegian design firm K8 and intended for off-grid use by the roughly a billion people that rely on hazardous kerosene lamps to provide lighting.

The SunBell case provides us with a context for the use of models, artifacts, and prototypes within a product design development process. While the end product is manufactured with conventional mass production techniques, such as plastic injection molding, AM tools takes up multiple roles throughout. It can be seen in the making of models as representations, through early conceptualizations of the lamp product, allowing the designers to negotiate functionality, usability, and shape. AM also acts as a tool for manufacturing, where it produces patterns for polyurethane (PUR) casting.

Marius Andresen, the founder of K8, saw the need to create a versatile sun-powered lamp that could meet a series of different lighting needs. While attending a workshop at the "Beyond Risør" event in Risør, in southern Norway, the participants were challenged with making lighting solutions based on various topics. K8 quickly decided on developing a lighting solution for replacing the hazardous kerosene lamps commonly found in less developed regions. Since its initial presentation at the lighting workshop in southern Norway, the initial mock-up has seen 4500 hours of research and development. The product is now sold commercially as well as through UNHCR incentives to countries such as the Philippines, Cameroon, Yemen, Sudan, Kenya, and Lebanon.

The entire process of developing the lamp, from early concepts to making functional prototypes and user-testing them, was conducted by K8. Much of this work is typically performed by the industrial designer, resulting in a set of functional and aesthetical models, accompanied by some guidelines and initial feedback. This is then presented to the client or manufacturer. This was not the case for K8, as the company itself had stakes in the project. After presenting the product concept at various design festivals, K8 found collaborating partners and created a dedicated company called Bright Products. With K8 as the developing entity behind Bright, investment for further development was found through private funding, corporate funding, as well as crowdfunding.

4.5.2 Initial Concept and Maturation of the SunBell Lamp

The SunBell project has been running since 2009, with the first batch of products delivered in 2014. Over a span of five years the product has gradually evolved in terms of composition and functionality.

The first concept of the SunBell (Fig. 4.6), composed out of simple IKEA parts, was made in April 2010 and featured an upside-down bowl with a penlight attached to it. The flexible goose neck, with the light in one end and a battery in the other, formed the principal structure of the product. In the following months, the basic concept with the flexible lighting source and the containing light-diffusing shade evolved into a basic operational model. Still using the IKEA hardware for the lighting and electronics, models gradually adapted to the new design. Using the in-house fused deposition modeling (FDM) 3D printer at K8, a new shape was made (Fig. 4.7) of the lampshade, allowing them to negotiate usability, functional components, and form giving.

Figure 4.6 The archetypal form of the lamp, made from IKEA components. Photo: K8.

"From an early stage we were making prototypes. From [the initial prototypes] we started to test out if we had enough space for the technological components, batteries, lighting, connectors and how the functionality fit with the overall concept," said Olivier Butstraen, who has been working with the SunBell project since the early conceptual stages. He is looking through a series of boxes containing prototypes and models, attempting to find the initial 3D-printed prototypes. This was not an easy find, considering all the broken pieces, artifacts taped together, and electronics partially assembled.

Figure 4.7 Initial negotiation of usability, functionality, and shape with initial rapid prototypes made with the in-house FDM 3D printer. Photo: K8.

A year after the initial workshop was held, K8 presented its first functional iteration of the SunBell lamp, as seen in Fig. 4.8. The initial product is entirely 3D-printed by their in-house FDM 3D printer. As the printer was only capable of creating rough models, they were painted and polished to simulate the materiality of injection-molded parts.

Figure 4.8 Functional prototypes of the SunBell presented at Beyond Risør in 2010. Photo: K8.

"At [the] prototype level we prioritized aesthetics and overall functionality [from a user perspective], more than considering the limitations of manufacturing. [The reason for] making the model was to show the concept and get feedback on it. So there was no need to solve technical details. In any way, the product was sanded down and painted," said Olivier. After being presented at the Beyond Risør festival and the establishment of the company Bright, the product was presented at events such as the Common Pitch in Canada, which the SunBell project won. The SunBell prototype gradually matured, although the overall shape and functionality were at this point set. "In many ways we worked ourselves inwards, as we started to get in contact with electronics experts, suppliers and manufacturers. But I am sure if you compared the silhouettes of all the models from day one to the mass-produced product you would see slight variations," remarked Olivier.

In 2012 a series of beta products were taken to Bangladesh, Kenya, and Tanzania for user testing. As eight models were planned for the trips, it was quickly decided to make a small series of products using PUR casting. Using the in-house 3D printer would simply be too much work. Instead, a series of silicone molds, one for each part of the lamp, was made using a high-resolution Objet printer to facilitate the making of molding tools for resin casting, as seen in

Fig. 4.9A. Once molded, the plastic parts were lacquered (Fig. 4.9B) in order to imitate the desired color of the product-to-be.

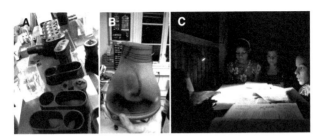

Figure 4.9 (A) Black polyurethane models that were molded for user testing. (B) After molding the models were lacquered in order to imitate the desired color of the product-to-be. (C) Field testing of the SunBell prototype in Bangladesh, Tanzania, and Kenya. Photo: K8.

The beta products proved to have two purposes—for gaining insight through field studies and for proving the marketability of the product. While the principal functionality and overall shape proved to be satisfactory in the field study, they gave the designers insight into potentials for improvement. Among others the integrated USB port would easily collect dust and could hinder the successful charging of a mobile phone. Observations such as this were then documented and would later help the designers in reiterating specific parts of the design before it went into mass production.

4.6   Design Representations and Multitypes in Product Design

In line with established development strategies, Capjon [8] introduced the concept of rapid multityping (Fig. 4.10), as opposed to conventional RP and rapid tooling, as a way of providing richer and more appropriate definitions to making representations in the development process. The approach is also contextualized in research by Killi, which develops a product-design-centered perspective and model on AM [24]. Capjon's vocabulary is heavily influenced by the emerging use of 3D printing as a strategy for constructing design representations and emphasizes the collaborative nature of designing new products.

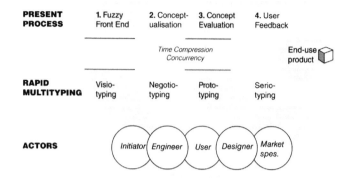

Figure 4.10 A redrawn model of multitypes. Redrawn illustration: William Kempton.

The rapid multityping models are distinct and refer to particular stages in the design process. Visiotyping is introduced as the process of developing visiotypes, initial mock-ups at the fuzzy front end of a design project. Negotiotyping deals with the actual conceptualization phase, with the relevant actors making a series of incremental negotiotypes for materializing mental imagery. The established notion of prototyping refers to the stage of concept evaluation, where prototypes are gradually introduced to other actors and stakeholders. Seriotyping emphasizes the functional and adaptive aspects of the user feedback phase, with seriotypes facilitating the event. No particular definition is provided for the process of manufacturing end-user products, although Capjon indicates the potentials of AM where conventional tooling is prohibitive or the possibilities of AM make it feasible.

The two models (Figs. 4.4 and 4.10) present contrasting views and taxonomies of VDRs used by designers. While the taxonomy of Pei et al. [34] offers a grid-like, structured, and highly detailed view of all the most commonly used forms of design representations, Capjon's model is less structured and does not provide a clear understanding of what the visio-, negotio-, proto-, and seriotypes might actually contain. As seen in the linear development process figure of Capjon (Fig. 4.10), no clear description of the representations is given, other than their being at different stages of the development process. Little emphasis is also given in Capjon's model concerning user feedback after the design development process is concluded. As a design process draws to an end, the design project is simply handed over to another party and eventually made into a product. The model also does not provide any clear transitionary information. The final development phase simply transitions over to the product space.

4.6.1 Multitypes in Rapid Prototyping

Multitypes may be applied with respect to RP (Fig. 4.11). They are useful in helping us understand the relationships between various stages in the design process. This can be seen in two cases. The first refers to the SunBell case mentioned above, concerning conventional manufacturing techniques and RP. The second case, with Pivot, refers to the role of 3D printing as it extends from a product development process.

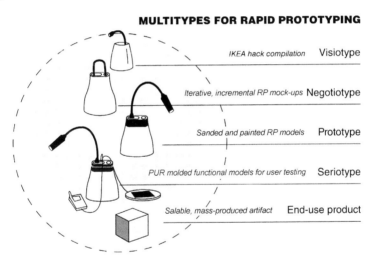

Figure 4.11 Multitypes in rapid prototyping. Illustration: William Kempton.

Case 1: SunBell

So far the SunBell case has provided us with a practical frame for the use of design representations within a design development process. The representations described in the SunBell case can be categorized within the multitypes as introduced by Capjon. The initial IKEA mock-up (Fig. 4.6), embodying the basic product functionality and composition, fits within the definition of a visiotype. The negotiotype, the early 3D-printed models using IKEA electronics, allowed the designers at K8 to refine the ergonomics of use and placement of hardware, as well as simulate the various lighting solutions. The conceptual model displayed at Beyond Risør embodies that of a prototype, as it represents the aesthetics and user functionality of the product-to-be. Lastly, the seriotype can be seen as the small-volume-production PUR lamps, which were used in field testing. From this case we can see that the multitypes are well defined and useful for the designers in their development process. As the actual product was intended for mass production by means of injection molding, the prototyping process culminated with the seriotype. Although minor alterations have since been made to the design of the lamp, (the lampshade has received an increased draft angle to enhance manufacturability), a redesign usually implies a reinvestment of costly manufacturing casting tools.

So far this case has described multityping for developing mass-produced artifacts, using 3D printing as a tool to facilitate RP. However, as 3D printing is entering a shift in paradigm, that is from RP to AM, we need to ask how multityping might be developed further. This may be done through reference to an experimental case that concerns the development of conceptual skiing grips by Pivot.

Case 2: Pivot

Pivot Industrial Design is a small design office located in Oslo, Norway. Its work spans from designing consumer and sporting goods to designing architectural installations. The company specializes in designing highly functional products and actively prototyping with 3D printers, laser cutters, and CNC mills in the development of them. "Everything is verified in our development process [before starting a manufacturing process]," exclaims Liam, one of the designers at Pivot.

The following case takes us through the story of developing a new generation of cross-country skiing poles. Pivot, in conjunction with Jørgen Weidemann Eriksen, found a need to improve the ski pole grip used by cross-country skiers for increasing the force output by 3%–4%. In principle, the grip tilts the user's hand so that more force is transformed into the pole, making each stroke more effective. The design of the ski pole handle started off as any usual product development done by Pivot. An internal brief was created and initial prototypes were made. Conceptual mock-ups were initially made in playdough and foam, later advancing to gypsum 3D-printed models for more accurate visual form finding and ergonomic verification (Fig. 4.12). At a certain stage in the development process it became apparent that for optimal use of the product, several sizes had to be made available in order to fit different hand sizes. At the same time, a new prototyping material was available to them, PA-11, a nylon material that surpassed the strength of other RP materials. While other RP prototypes proved short lived, the selective laser sintering (SLS)-produced PA-11 had a superior material structure, enabling the designers to consider the possibility of using AM as an actual production technique.

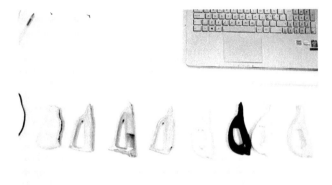

Figure 4.12 Incremental development (left to right) through a series of models from the DF1 ski pole grip. Photo: Pivot Design.

According to Liam, "This basically allowed us to start thinking about completely new ways of encountering production limitations. No longer did we need to limit the design to even wall-thicknesses, or draft angles, [common considerations when designing for injection molding]." From the very beginning, the product was focused on a higher, more professional segment. Current ski pole grips are usually made of a combination of injection-molded plastics, as well as organic materials such as cork and leather. The 3D-printed PA-11 was considered as a replacement for all the existing materials. However, unlike the existing ski pole grips, which were designed to fit a larger variety of hand sizes, the new grip would have to come in a variety of sizes to accommodate the ergonomic fit of different hand sizes.

"In the end the product never materialized for lack of external investment, but this could certainly change within a short time," Liam reflects. Although the product was never taken further than a conceptual stage, the designers at Pivot found a space for AM as a feasible method of production, with the potential for incorporating user-specific ergonomic fit.

4.7 Multityping in Additive Manufacturing

4.7.1 Popular yet Professional?

The two cases mentioned above, the SunBell lamp and Pivot's ski pole grip, further a discussion on the implications of AM. While the first case illustrates development practices that integrate RP for making various forms of design representations, the second case advances the idea of extending RP equipment into a mode of production in order for Pivot to adapt to the end product to user-specific ergonomics.

Through online media libraries such as Netflix and Spotify, unfiltered access to culture and media content is being provided to a realm of users. Whereas traditional broadcasting, such as radio and TV channels, have been able to provide selected, curated content to the millions of viewers, the distribution of media through the Internet turns the ratio around: millions of varieties of music and film are available to each user. This scattering of the mass market to many niche markets is what Chris Anderson [2] refers to as the "Long Tail." The notion can also extend to physical contents, as online stores such as Amazon.com make new products and contents available to masses of people. Concerning the Long Tail of Things, 3D printing is envisioned as a distributed mechanism for manufacturing, with the desktop factory printer turning "bits into atoms in your own home" [2]. Although desktop 3D printing technologies are becoming quicker and more reliable, bringing these tools into domestic environments has its own implications. Aside from being restrictive in terms of size and choice of material (most desktop inkjet printer owners only have printers that support A4-size papers), the accompanying software tools for creating 3D models often require extensive practice. Many of these challenges can also be addressed through increased online collaboration. Online sharing platforms such as Thingiverse (thingiverse.com) already allow enthusiasts to share and codevelop products, while online video repositories such as YouTube (youtube. com) have become popular venues for online learning. The space in which the 3D printer is situated can also be virtual. Of note are popular 3D printing services such as Shapeways (shapeways.com) and 3DHubs (3dhubs.com), which give those who didn't previously own their own desktop fabricator distributed access to 3D printing services.

On consumerism and product innovation, von Hippel [20] discusses various modes of increased individualization. He clearly identifies a gap between users' diverse needs and the homogeneous offerings of mass-produced artifacts. This is partially identified though what he describes as lead user innovation, where engaged users, who are experiencing needs that will later be experienced by other users, actively and freely participate in organized innovation processes. Hippel [20] extends this into a "toolkit for user innovation and custom design," whose goal is to enable nonspecialist users to partake in the design of high-quality, manufacturable products that are customized to their specific needs.

4.7.2 Integrating AM in Product Design

From a design view we see that AM is in fact not just a replacement of current manufacturing techniques but the embodiment of a new way of thinking product design. Here, the processes of business development, engineering design, marketing, and operations are increasingly set closer together—as digital, physical, or hybrid products.

So far in this chapter I have attempted to create a holistic vision of how 3D printing affects the process of developing products. While the SunBell case related to a conventional paradigm of multityping, where the resulting models and artifacts were merely rapidly prototyped representations of a mass-produced artifact, the DF1 ski pole grip and Durr watch come to represent a different form of multityping. In design projects where conventional manufacturing is used, there is a large shift between latter stages of prototyping, seriotyping, and making the actual product, as investment costs for manufacturing tools are prohibitively high. However, in an AM paradigm, the step between the seriotype and a usable or commercial product can be seen as more fluid. Although new manufacturing tools may be used, or a supply chain needs be created, the investments costs for AM are considerably lower than conventional tool-based manufacturing. In addition, the customizable and enhanced aspects of AM products, such as products that take into account the materiality of AM, can be harnessed by online, digital services, such as digital-physical toolkits [19].

4.7.3   Toward the Releasetype

Unlike Knutsen's hybrid products [25], which are networked, connected devices with dynamic functions, the releasetype is dynamic in its digital form but fixated as it is physically reproduced. While the releasetype is available and accessible to people, it has a notion of temporality, always subject to change—in shape, content, and materiality. The releasetype is a result of a series of design representations made in a design development process, known as multitypes [8]. While other types—such as prototypes—are representations and models for internal product development, the releasetype lives in the open world, accessible to its end users. To further my argument of fluid, developmental transitions between material artifacts, I will first elaborate on the developmental practices of design and prototyping.

In AM products the transition from functional, usable seriotypes to the released product should be interrupted by a new classification of type—the releasetype. A releasetype is a digital/physical hybrid artifact consisting of a design framework and a user interface. As opposed to conventional product design where a typical design task is the shaping of a single artifact, the new design task becomes the forming of the releasetype solution space and interface. The releasetype is still the result of a systematic design development, incrementally matured through stages of visiotyping, negotiotyping, prototyping, and seriotyping. However, as a digital/physical product hybrid, the releasetype is tangible to its users primarily through an interface, where it can be extracted and fixed into physical space.

In Fig. 4.13, the releasetype is placed at the end of a succession of multitypes, to emphasize the iterative nature of design development. Whereas the preceding, physical multitypes can be seen as following a river fall cycle between the design development and stakeholders, the releasetype feedback loop is more immediate, converging between a physical and digital space.

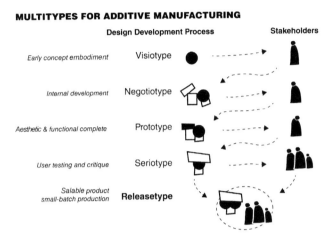

Figure 4.13 Multitypes for additive manufacturing. Illustration: William Kempton.

4.8 Conclusions

This chapter investigates the emerging use of digital fabrication from a viewpoint of product design practice. The question in focus is, how is the emerging, expert use of digital fabrication providing new models for product conceptualization? 3D printing, in particular, has been primarily seen as a prototyping tool, giving it the name "rapid prototyping." As a collection of technologies for reproducing physical artifacts, based on a digital blueprint [21], the emerging trend of 3D printing is gradually moving toward production of usable artifacts. This is due to the gradual improvement of material properties, affordability, and speed [13].

Earlier, I looked into how the Durr wristwatch, the SunBell lamp, and the ski pole grip function as some examples of how 3D printing is changing the landscape of additive product development. As one unbuckles the Durr watch, it is possible to see some of the new potentials and relationships between digital design space and product design. To account for such spaces for design in digital fabrication, I proposed the concept of releasetypes. This, I suggested, is a complement to the established use of prototypes in design inquiry and production. The introduction of releasetypes calls for a repositioning of the traditional sense of physical products into new kinds of digital/physical hybrids. However, unlike networked, connected hybrid products, the releasetypes are themselves digital design frameworks; they are manifested through physical materials when digitally fabricated. The introduction of the releasetype as a physical/digital hybrid product spurs on new discussions as to how it will change the role of design.

So why is this relevant for the designer? The purpose of the designer may be understood as expanding from balancing the broad variety of uses, use areas, and ways of using products to facilitating an increasingly cocreative space where both users and designers are able to work together. While 3D printing has been a part of design development for several decades, there is an increasing momentum in the development of AM and 3D printing, as it passes beyond professional use and into public, transdisciplinary access.

References

1. Abras, C., Maloney-Krichmar, D., and Preece, J. (2004). User-centered design. *Bainbridge, W. Encyclopedia of Human-Computer Interaction.* Thousand Oaks, CA: Sage Publications, *37*(4), 445–456.

2. Anderson, C. (2007). *The Long Tail: How Endless Choice is Creating Unlimited Demand.* London: Random House.

3. Belliveau, P., Griffin, A., and Somermeyer, S. (2004). *The PDMA ToolBook 1 for New Product Development.* Hoboken, NJ: John Wiley and Sons.

4. Best 3D Printer Guide 2017 (2016). Retrieved January 23, 2017, from https://www.3dhubs.com/best-3d-printer-guide

5. Bowen, S. J. (2009). A critical artefact methodology: Using provocative conceptual designs to foster human-centred innovation (Doctoral dissertation). Sheffield Hallam University. Retrieved from http://shura.shu.ac.uk/3216/

6. Bowyer, A. (2004). Wealth without money. Retrieved from http://reprap.org/wiki/BackgroundPage

7. Buchenau, M., and Suri, J. F. (2000). Experience prototyping. In *Proceedings of the 3rd Conference on Designing Interactive Systems:*

Processes, Practices, Methods, and Techniques. New York, NY, USA: ACM, pp. 424–433. http://doi.org/10.1145/347642.347802

8. Capjon, J. (2004). Trial-and-error-based innovation: catalysing shared engagement in design conceptualisation (Doctoral dissertation). The Oslo School of Architecture and Design, Oslo, Norway.

9. Cooper, R. G. (2011). *Winning at New Products: Creating Value Through Innovation.* New York: Basic Books.

10. Cross, N. (2006). *Designerly Ways of Knowing.* London: Springer.

11. Cross, N. (2008). *Engineering Design Methods: Strategies for Product Design.* Chichester: John Wiley & Sons.

12. Crump, S. S. (1992). Apparatus and method for creating three-dimensional objects. Retrieved June 9, 1992, from http://www.google.com/patents/US5121329

13. Doubrovski, Z., Verlinden, J. C., and Geraedts, J. M. (2011). Optimal design for additive manufacturing: opportunities and challenges. In *ASME 2011 International Design Engineering Technical Conferences and Computers and Information in Engineering Conference.* American Society of Mechanical Engineers, pp. 635–646.

14. Eisenberg, M. (2013). 3D printing for children: what to build next? *International Journal of Child-Computer Interaction*, **1**(1), 7–13.

15. Eppinger, S., and Ulrich, K. (2015). *Product Design and Development.* McGraw-Hill Higher Education.

16. Gerber, E. M., and Martin, C. K. (2012). Supporting creativity within web-based self-services. *International Journal of Design*, **6**(1), 85–100.

17. Gibson, I., Rosen, D. W., and Stucker, B. (2010). *Additive Manufacturing Technologies.* New York: Springer.

18. Grieser, F. (2016). 20 best 3D printing software tools (most are free). Retrieved March 21, 2016, from https://all3dp.com/best-3d-printing-software-tools/

19. Hermans, G. (2014). Investigating the unexplored possibilities of digital–physical toolkits in lay design. *International Journal of Design*, **8**(2), 15–28.

20. Hippel, E. von (2005). *Democratizing Innovation.* Cambridge, MA: MIT Press.

21. Hopkinson, N., Hague, R., and Dickens, P. (2006). *Rapid Manufacturing: An Industrial Revolution for the Digital Age.* Chichester: John Wiley & Sons.

22. Houde, S., and Hill, C. (1997). What do prototypes prototype. *Handbook of Human-Computer Interaction*, **2**, 367–381.

23. Hull, C. W. (1986). Apparatus for production of three-dimensional objects by stereolithography. Retrieved March 11, 1986, from http://www.google.com/patents/US4575330

24. Killi, S. (2013). Designing for additive manufacturing: perspectives from product design (Doctoral dissertation). Oslo School of Architecture and Design, Oslo.

25. Knutsen, J. (2014). Uprooting products of the networked city. *International Journal of Design*, **8**(1), 127–142.

26. Knutsen, J. (2015). *Products of the networked city: exploring and revealing the materials of networked and computational infrastructures* (Doctoral dissertation). Oslo School of Architecture and Design, Oslo.

27. Koen, P., et al. (2001). Providing clarity and a common language to the 'fuzzy front end'. *Research-Technology Management*, **44**(2), 46–55.

28. Kumar, V. (2012). *101 Design Methods: A Structured Approach for Driving Innovation in Your Organization*. Hoboken, NJ: John Wiley and Sons.

29. Kvale, S. (1996). *InterViews. An introduction to Qualitative Research Writing*. Thousand Oaks, CA: Sage Publications.

30. Lawson, B. (2006). *How Designers Think: The Design Process Demystified*. London: Elsevier Architectural Press.

31. Liang, L.-H., and Paddison, L. (2016). Could 3D printing help tackle poverty and plastic waste? *The Guardian*. Retrieved November 6, 2016, from https://www.theguardian.com/sustainable-business/2016/nov/06/3d-printing-plastic-waste-poverty-development-protoprint-reflow-techfortrade

32. Lim, Y.-K., Stolterman, E., and Tenenberg, J. (2008). The anatomy of prototypes: prototypes as filters, prototypes as manifestations of design ideas. *ACM Transactions on Computer-Human Interaction*, **15**(2), 7.1–7.27.

33. Milton, A., and Rodgers, P. (2013). *Research Methods for Product Design* (1st ed.). London: Laurence King.

34. Pei, E., Campbell, I., and Evans, M. (2011). A taxonomic classification of visual design representations used by industrial designers and engineering designers. *The Design Journal*, **14**(1), 64–91.

35. Pham, D., and Gault, R. (1998). A comparison of rapid prototyping technologies. *International Journal of Machine Tools and Manufacture*, **38**(10), 1257–1287.

36. Pine, B. J. (1999). *Mass Customization: The New Frontier in Business Competition* (New ed.). Boston, Mass.; London: Harvard Business Review Press.

37. Plowman, T. (2003). Ethnography and critical design practice. *Design Research: Methods and Perspectives*, 30–38.

38. Ratto, M. (2011). Critical making: conceptual and material studies in technology and social life. *The Information Society*, **27**(4), 252–260.

39. Ratto, M., and Ree, R. (2012). Materializing information: 3D printing and social change. *First Monday*, **17**(7). Retrieved from http://journals. uic.edu/ojs/index.php/fm/article/view/3968

40. Ree, R. (2011). 3D printing: convergences, frictions, fluidity (Doctoral dissertation). University of Toronto, Toronto, Canada.

41. Robins, L. M., and Kanowski, P. J. (2008). PhD by publication: a student's perspective. *Journal of Research Practice*, **4**(2), 3.

42. Runberger, J. (2008). Architectural prototypes: modes of design development and architectural practice (Licentiate thesis). KTH School of Architecture and the Built Environment, Stockholm, Sweden.

43. Salvador, T., Bell, G., and Anderson, K. (1999). Design ethnography. *Design Management Journal (Former Series)*, **10**(4), 35–41.

44. Schön, D. A. (1983). *The Reflective Practitioner: How Professionals Think in Action* (1st ed.). New York: Basic Books.

45. Sennett, R. (2008). *The Craftsman*. New Haven, CT: Yale University Press.

46. Sevaldson, B. (2011). GIGA-mapping: visualisation for complexity and systems thinking in design. *Nordes*, **0**(4). Retrieved from http://www. nordes.org/opj/index.php/n13/article/view/104

47. Lurås, S., Lützhöft, M., and Sevaldson, B. (2015). Meeting the complex and unfamiliar: lessons from design in the offshore industry. *International Journal of Design*, **9**(2), 141–154.

48. Takeuchi, H., and Nonaka, I. (1986). The new new product development game. *Harvard Business Review*, **64**(1), 137–147.

49. The printed world (2011). The economist. Retrieved February 10, 2011, from http://www.economist.com/node/18114221

50. Tovey, M. (1989). Drawing and CAD in industrial design. *Design Studies*, **10**(1), 24–39.

51. Yin, R. K. (2013). *Case Study Research: Design and Methods: Design and Methods*. Thousand Oaks, CA: Sage.

Chapter 5

Visual 3D Form in the Context of Additive Manufacturing

Nina Bjørnstad[a] and Andrew Morrison[b]
[a]*Institute for Design, Oslo School of Architecture and Design, Oslo, Norway*
[b]*Centre for Design Research, Oslo School of Architecture and Design, Oslo, Norway*
nina.bjornstad@aho.no; andrew.morrison@aho.no

5.1   Introduction

Digital fabrication, maker spaces, fabrication laboratories (fab labs), and 3D printing have grown strongly in the past five years. Bold claims have been made about the transformatory power of these technologies, ranging from their properties in terms of producing rapid, additive, and iterative versions in a design process to their contribution to emerging practices of mechanical, automated, and tool-driven production. Rather surprisingly, little attention has been given to one of the key features of additive manufacturing (AM): the shaping of 3D form. This may lie in the development of the name AM itself.

Earlier called rapid prototyping (RP), "additive manufacturing" became the preferred term to replace the speedy, iterative, and

Additive Manufacturing: Design, Methods, and Processes
Edited by Steinar Killi
Copyright © 2017 Pan Stanford Publishing Pte. Ltd.
ISBN 978-981-4774-16-1 (Hardcover), 978-1-315-19658-9 (eBook)
www.panstanford.com

prototyping character of the technology that allows for the production of 3D objects, components, and artifacts. This production takes place in a dynamic process of shifting between embodied knowledge and understanding of 3D physical artifacts and the shaping of 3D forms via digital design tools on-screen and their transfer via 3D printing technologies from 2D digital renderings to tangible, tactile, and physical volumes and products. The additive character of this process lies in the potential to shift backward and forward between the conceptual and the physical and to potentially work within an expanded space of 3D form.

However, much of the research literature centers on engineering, manufacturing, and technological aspects of the technologies (e.g., Ref. [8]), such as articles in the journal *Virtual and Physical Prototyping.* Less attention is given to design-related aspects of what it is that makes AM a tool for designing products in an age of digital design and in relation to the design affordances of 3D printing. One central issue is the extent to which approaches to 3D visual form are maintained or altered through the uptake of digital fabrication technologies. This chapter addresses this issue with respect to one of the major design-based analytical models on visual 3D form and its application as a resource for both design and interpretation.

In taking up this model, we consider the impact of shifts between digital design and 3D physical rendering in teaching product design (PD). To what extent is this model applicable for working with 3D visual form in relation to AM technologies? What influence do the tools have on the model and its practices in form-giving pedagogy and processes? Do the affordances of AM—rapidity, malleability, iterativeness, etc.—stretch or change our perceptions and processes of 3D visual form development? Further, what might prior and emergent understanding of working with 3D form in terms of aesthetics and form giving contribute to a richer and PD-centered view of AM.

5.2 Aesthetics

As a representative for aesthetical aspects of 3D form, the following concepts are first and foremost a reminder to support artistically compositional training and the need for modeling experience behind successful design. Even though it's not as naive as believing the earth is flat, describing 3D objects by their 2D look-alike is common. Colleagues carelessly describe 3D forms with 2D terminology, though this should be a core competence. A cone may be described as a triangle; the third dimension is forgotten or left out. A triangular prism is often called a "Toblerone form." People commonly rely on similes and metaphors to refer to visual 3D forms.

The abstract formalistic language from the early twentieth century still prevails in art and design today. This applies to professional design practice and to design education. Akner-Koler [3] argues that this formalistic language has its genesis in scientific thinking together with artistic sculptural development. Today, toward the end of the second decade of the twenty-first century, principles bound in geometrical law are widely used in industrial design products. Let's just consider an iconic car. Seen from the side, the compound form of a Volkswagen (VW) Beetle is possible to frame or inscribe in a circle. Front on, the VW sign is positioned exactly in the center of a square framing the car's silhouette. The name "Beetle," though, is a metaphor that clearly communicates form. When it comes to communication with design students and users we want to point out the importance of references to well-known forms. However, formalistic descriptions lack cultural references; they might be perceived as elements of a professional language and thereby be rather alienating. We argue that designers need both these elements in order to understand and to communicate their work to others. This applies equally to teasing our relations between PD and AM, that is, formally, pedagogically, and analytically.

In his doctoral thesis about the structure of projects relating to RP, Capjon [5] wrote that "aesthetic reasoning is possible through operational actions on physical forms." For Capjon, "RP as a development tool is generally applied, as the name indicates, for prototyping of more or less finished concepts" [5]. What then of the exploration of abstract 3D visual forms in AM?

This chapter addresses abstract concept creation in the early phase of design and its potential for exploring relations between form, aesthetics, and AM. This aligns with what Capjon [5] referred to as "an aesthetical approach to design." However, a piece of additively manufactured design cannot be realized without engagement with tools and processes concerning digitalization, production, and distribution. This also extends to the iterative and dynamic processes of prototyping. The repeated techniques of such RP in the development of design artifacts, as Capjon suggests, demand digital skills. The merger of suitable software and digital literacies relating to design is central to how an aesthetic view may be elaborated. This applies conceptually in the early phases of design but also additively as prototypes are revised and replaced. This is not simply a matter that before printing the 3D form it is necessary for it to be transformed into bits and bytes (see Chapters 3 and 4, Sections 3.1.1.2 and 4.5.1).

What is needed is attention to aesthetic 3D form from both the physical and the digital, that is prior to, during, and after the design and the printing of artifacts. A dynamic ought then to ensure that the designer is allowed to reform or recalibrate his or her digital settings to realize new 3D-printed objects that are again assessed and reframed as needed. It is surprising that virtually no explicitly formal reference has yet been made to aesthetic form generation and elaboration in the research literature (see, for example, Ref. [11]) or many handbooks relating to AM. Aesthetics is taken up where AM is applied in electronic art (e.g., Ref. [9]), but the focus is typically not on form finding, transformation, and dynamic processes of aesthetic development for 3D-printed artifacts. This is the domain of PD and may be teased out by referring below to abstract approaches, to models of 3D visual form, and to related professional terms as we present in detail below. Adopting the view of imaginary deconstruction (looking at form like an x-ray) allows us to retrieve the skeleton of a suggested design. A breakdown into basic geometry is necessary in order to digitalize the form. This is linked to the method of abstraction of this chapter, where basic geometry is used to create a multitude of clay models as the starting point.

Practical skills in creation of 3D forms, backed up with professional terminology, are essential in design work. Designers need geometrical knowledge and training in creating, constructing, and finally describing form. However, neither students nor professionals are well-enough trained in the vocabulary. It is this theme around which we hope to contribute insights regarding 3D form and AM. Earlier industrial design, now generally labeled "product design," provides core models upon which to not only build but also extend experimentation and design work with AM tools and technologies.

Clearly established since the 1930s, industrial design provides the foundational model for many schools of design. These too were premised on earlier traditions in use in art and craft schools around the world. There are differences in strategy on the part of art and design schools when it comes to conceptualizing and enacting strategies for meeting user demands. One matter in common is that of the creation of aesthetical diversity. This we take up later, in terms of contextualization, in Section 5.4.

5.3 Design, Action, and Profession

The development of teams has different points of view and intentions. Where then does the idea for different forms in design come from? Earlier we pointed out the form giving [2] that takes place in the earlier part of the product development process (see Chapter 3, inside the multiple stakeholder product development process, Fig. 3.3). To develop concepts, designers need to work out a balance between using codesign processes and studio time on their own. A design task might have conventional answers, but designers are trained to "go abstract" in the early process and experiment with form and space. We suggest it is through such a tactic of following basic form phenomena that it is possible to come up with a greater diversity of designs and that this mode of generation of designs is still pertinent and needs to not be forgotten when designers take up digital tools and their claimed affordances for transforming 3D production.

With the possibilities inherent in AM, the variation of form is potentially infinite, though there are limitations (see Chapter 2). What we aim for is physical form; designers and students have to relate to the laws of nature. The system called the evolution of form (EoF) model that we describe later has been developed for higher design education and is accessible online [2]. EoF aims to structure understanding and ways of entering into exploration with form that are accessible for industrial designers. We argue this is a liberating model for further learning activity in the ongoing development of AM that aids well its design-oriented explorations and application.

We aim to demonstrate how it may function and an analytical way for designers to generate multiple forms, whether one is fresh out of school or an experienced designer.

5.4 Ideals and Origin

In 1934, sculptor Rowena Reed and painter Alexander Kostellow founded the first industrial program in the United States at the Pratt Institute, New York City [1]. Both of these innovators of this later influential school came from the arts. The systemic approach for understanding form in design they established remains in use today. Reed met Kostellow at the Kansas City Art Institute in the 1920s, where Kostellow was an instructor in painting. Kostellow contributed his pictorial structure and experience in organizing axes on the canvas, and Reed experimented with the visual organization in 3D.

The Russian American sculptor Alexander Archipenko subsequently introduced space studies as well as the constructivist perspective on which Reed based her teaching. Reed further included the spontaneity central to abstract painting within developmental methods of sculpturing without reducing 3D complexities [1]. The foundation program became the prototype for American schools and later European ones (Elements of Design [EOD] [7]). It was the coherence of this educational program that distinguished it from the modernistic approach developed at the Bauhaus and through the Euro/Russian Constructivist Movement [3]. Akner-Koler [1, 2] argues that Kostellow and Reed were not engulfed in the Russian Constructivist Movement. Both of them originated in the arts, and Reed's 3D organic methodology was based on abstraction of the figure [1], influenced by Archipenko.

In his article "Form Follows WHAT?" Jan Michl [15] considers ideals in design education and discusses the modernist notion of function as a carte blanche. He discusses, among other things, whether designers and educators unintentionally continue a set of stylistic ideals from the midtwentieth century by not addressing users and their tastes, wishes, or dreams. Writing about a 3D visual method without including color or material reveals our close link to the modernistic ideals, as Michl argues.

Design has to be conscious about the operation of subjective criteria concerning taste and culture; this is critical for how products are interpreted. It is a profession where subjective criteria, taste, and culture come into consideration both in product and surrounding communication. An understanding of the taste of the users guides the development team toward a negotiated aesthetics optimized for the context. Creating and communicating complex form benefit from a multitude of models.

The early product sketches are often abstracted or simplified; you don't bring in the on/off button in the first sketches. This training in abstract form is preparing the design students for a multisolution attitude and comes before and during involvement of the user, making it possible to have others evaluate the objects.

5.5 Relevance of the Historic Model for Tomorrow's Form Givers

The EoF model shown below is an example on how form has been developed further in the theoretical, practical, and pedagogical work of Cheryl Akner-Koler. Form experiments are not intended to serve as the end product. The relevance of form exploration is to give designers an enhanced experience and a pool of form ideas to make use of in design projects. Through this approach and its pedagogical orientation, we find ways of generating form for students, professional designers, makers, and engineers. Even though the technology is immediate and at hand, it is the attention to the inherent potentials of the form that is important, their iteration and revision. Therefore, the EoF model is relevant because of its qualities concerning the facilitation of how form may be explored. Matching the low-tech of the clay with the high-tech AM makes the method relevant for quick iteration and small-scale production.

Reed and Kostellow did not document their teaching methods themselves. However, in 2002 a former student of Reed's, Gail Greet Hannah, drew up a bibliography through voice recordings from Reed's teaching sessions and by interviewing former Pratt students [7]. The Pratt/Kostellow/Reed heritage of teaching principles, model, and methods has continued to developed through generations of designers teaching design students. The curriculum has developed, and new form phenomena are added according to current technologies and trends. Additions might have been made due to the possibilities within plasticity of casting methods, from material properties or contemporary production technology. Transitional forms such as the operation "merge" are found in forms made for injection molding or most other productions methods, taking over the simple bend-and-weld actions.

5.6 The Evolution of Form Model

The EoF model by Cheryl Akner-Koler [1, 2] shown in Fig. 5.1 is a systemic form-mapping system applied to the creation of form. It goes from geometrical to organic form, which shifts from simple to complex through relevant form phenomena that are sorted on a bipolar spectrum. Akner-Koler's empirical data and research was developed at Konstfack (University College of Art, Craft and Design) in Sweden. Her system refers to the form and space experiments that Reed and Kostellow taught at the Pratt Institute. Exercises from applying and exploring this model serve to help future designers to understand diversity as well as to apply better terminology and communication for describing and interpreting form.

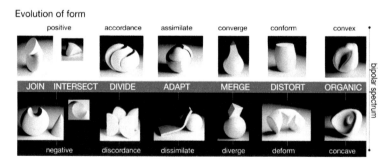

Figure 5.1 Evolution of form (EoF) model by Cheryl Akner-Koler (1994, 46 ff).

Terminology for describing a product can vary in character. "It's tall and elegant" and "It's a vertical stretched cone" point to two different ways to communicate. "Elegant" is subjective and diffused, and "stretched cone" is a more objective description. Design is usually developed by cross-disciplinary teams responsible for market, economics, or engineering, hence the importance of visualizations and objective terminology together with seeing the product in its right context. Following Akner-Koler, it is through abstract forms that new ideas emerge and solve expressions and needs in an experimental way. The shapes generated are initially not transformed further into a product but will point onward as it were through their affordances for possible functions. The relevance of using EoF is that it generates basic knowledge in form, operations, and systems for changing form that, in turn, supports the development of composition more fully.

5.6.1 Why Clay?

Looking at 3D form in an AM context we see an immediate dependence on computer-aided design (CAD) and associated digital tools. Akner-Koler acknowledges the guiding effect the digital toolbox has on virtual form. A corresponding set of skills is clearly needed. According to Malcolm McCullough [14] we prefer digital tools when they have an affinity to a conventional process. A well-known physical finishing work called fillets and chamfer—terms from the wood workshop—is to be found in computer programs. More specialized are splines. A spline is a flexible strip used in drawing curved lines, especially on hulls. The thin piece of wood, rubber, or metal is the origin of the digital curve tool by the same name.

Both software and hardware make creating form less intuitive; the tools add a hindrance between the hands and the mass. For McCullough [14], "Much life of the hands are a form of knowledge: not a linguistic or symbolic knowledge . . . but something based more on concrete action, such as sculpturing plaster or clay." Strangely enough, within interaction design, in the context of form creation, input technology still lacks haptic interfaces. Without the exploratory and manipulative aspects of touch as form, the symbols and graphics remain purely at the level of the visual interface (Fig. 5.2). Movements of the computer pointer together with graphics convey "direct action" [13], but this is without close relation to the body. New readings of gestures seem promising to the development of the input technology though these needs would benefit from both greater attention to their form and aesthetical aspects in the context of digital fabrication (e.g., Ref. [19]).

Figure 5.2   Shaping forms with clay. Photo: Maria Karlsen.

So, with developments in software tools and advances in embodied interaction, why do we work in clay and why directly in three dimensions? When working in physical three dimensions, emphasis is put on volume, space, depth, and all-roundness [2], which complements working from specific fixed viewpoints stressed in 2D works. To fully understand proportions and scale in relation to the human body is reason enough to use clay for 3D sketches. Freedom to quickly add or subtract material lowers the threshold for change. The level of detailing is low in comparison with in digital programs. This suits early sketching. In addition, rough surface treatment in early sketches also draws attention to the inner structure of form rather than the surface and silhouette.

Software resembles real-life production methods. The deconstruction of clay forms refers to similar production methods: when you "cut" something in the digital tools you refer to a virtual form of a physical cut. There is a physical manifestation in a digital mode. One teaches form from the physical and digital points of view together then but the digital manifestation makes it possible to devise quicker iterations. There are always exceptions: in the virtual space one works with features such as tweaking that cannot be done in the physical space: sometimes it results in new physical phenomena that cannot be worked out in the physical spaces. The EoF model allows us to look into both these processes and phenomenon. Further consideration of form is, therefore, central to our ongoing understanding of relations between the physical and digital as AM continues to expand its reach. Yet there is still need to disentangle some key terms that the field of AM would benefit from using more precisely.

5.6.2   A Close-Up on Form

In 1974, Arnhem wrote that "form always go beyond the practical functions of things by finding in their shape the visual qualities of roundness or sharpness, strength or fragility, harmony or discord" [4]. Further, Ching [6] describes form and shape as follows: "Form is the primary identifying characteristic of a volume. It's determined by the shape and interrelationships of the planes that describe the boundaries of the volume."

In his book *Principles of Form and Design* Wucius Wong [20] developed a visual form of grammar and defined constructional elements. Vertex, edge, and faces are examples and can help in defining volumes. According to Wong, for example, a cube has 8 vertices, 12 edges, and 6 faces. In terms of design processes and development, we are able to discuss these properties in detail since they are captured in the clay models without user function as in products.

In Wong's structure we see a skeleton, or in a formalistic manner an inner axis, and its construction and organization. The cone is a single curved surface meeting a circular plane, where its vertical axis rises from the center of the circular plane. When we execute the "distortion task" (the sixth stage in the EoF model) we work with or against the structure. Basic geometrical forms have their own characteristics. As the tetrahedron consists of four triangles, or a cylinder of two circular planes and a rectangular single curved side, we can deconstruct them into components.

To fully explore the form expressions it is crucial to work in physical material freely (Fig. 5.3). In the following samples, "form phenomena" are explained and exemplified. We isolate particular form phenomena in order to look specifically at certain effects.

We exclude color and transparency in order to focus on structural composition, but maintain some sense of texture.

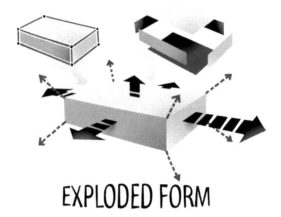

EXPLODED FORM

Figure 5.3 Exploded form. Imaginary deconstruction of the volumes into constituent parts. Drawing: Gjermund Bohne.

The following sections show research conducted by Nina Bjørnstad in further developing a systematic abstract exploration in today's educational context. The first-year students at the Institute of Design at the Oslo School of Architecture and Design (AHO) made clay models during the spring term in 2016. Nina's pedagogical teaching methods are derived from her experience as a student in courses taught by Professor Cheryl Akner-Koler at Konstfack in the early 1990s. At that time, Konstfack offered a master's of fine art in industrial design. The school also offered a foundation and artistic course that applied color and form training from a fine arts perspective, taught by artists who tutored in groups and applied "learning by doing" processes. This "color and form" program built and connected knowledge and experience as well as intuition and skills. Nina has taken this approach across to AHO and now is engaged in introducing it further to AM settings. Analytically to our knowledge this is not being done in other product design and AM initiatives where the focus may still be, for example, on materials investigation or product development sequences.

As part of the spring 2016 course, at AHO, in order to visualize the different aspects of the EoF model, form phenomena were investigated and visualized. The examples we include here draw from the EoF model and from referenced literature by Akner-Koler.

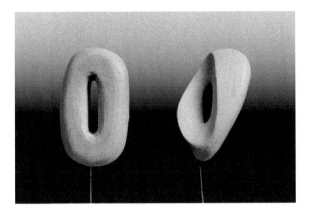

Figure 5.4 Distortion of a torus. The picture shows two different interpretations of distortions of a torus. The first torus is evenly crushed from two opposite sides. The second one is deformed as it was thrown around the center of the torus. The interpretation limits the deformation as the torus was made of elastic material. The imaginative movements are frozen in the clay and resemble a disciplined interpretation of dynamics within the mass. The concave outline from the center hole of the torus is accentuated in a catapulted edge out from center. Model: Marie Eckhoff Owing; picture: Maria Karlsen.

5.6.3 Distorted Forms

Internal or external forces are at work here. The students start by modeling their own basic forms (sphere, cube, cone, prism) and proportions (extensional, massive, superficial). They then imagine an external or internal structural force affecting the axis that is working inside the mass. In clay, we can build form inside out (additive shaping) or we start with a solid block (subtractive shaping). With subtractive shaping it may be possible to cut away or squeeze, while milder distortions could involve pushing or bending the object (Fig. 5.4). If we go one step further, we can imagine the form imploding or exploding into pieces. The latter is a kind of interpreted effect modeled into clay, rather than a direct force on the volume.

The force can be narrow or wide or soft or strong. Internal forces can twist, implode, or explode the form with which they started.

Where the force has gone through a form, it pierces the form and introduces new properties, such as holes. Such deformations provide new complexity to the form with which we started. The challenge is not only to imagine but also to transform the result and to be able to shape it in a wide range of ways.

The immediate distortions are sometimes used even though they can tend to be somewhat excessive for a design project. The discussion that needs to occur here is one that focuses around understanding a spectrum of forces working with the properties in the form. There is a shift from "conform to deform." Here, there are more subtle effects, not exploded or fragments, that we are looking for. A typical feature in design objects is concave surfaces to signify grip. This form phenomenon can be more expensive to enact in many methods of production due to the fact that asymmetry leads to expensive tools or due to complicated tools in form blowing. Yet, distortion of form opens up possibilities to an almost infinite range of shapes, suitable for freeform production and for exploring form elasticity in custom-made products.

The examples from the student workshop are all made by hand and are not developed into a product. But how could this be done? If we again take a look at the drinking container task presented in Chapter 3, one of the outcomes was a cup that was deformed in different ways, (see Fig. 5.5). Distortion, you will recall, is one of the principles presented in the EoF model. These containers also show another effect arising from the digitizing process. When a digital model is prepared for 3D printing, the digital file is very often translated to what is called the .stl (or stereolithography) format. This format represents form by flat triangles. If the triangles are very small, that is, millions of them, we could perceive, for instance, a sphere to actually be a sphere and not the jagged surface of thousands or millions of triangles. This may be used in the design process. The deformed containers in Fig. 5.5 were originally a smooth, rounded, form. By running the translation to stl with a low density of triangles, thereby reducing the resolution, we get the effect of surfaces seen on the containers. So, not only could the containers be deformed in many ways, the density or resolution of the file to be printed could result in a second effect. This latter action provides the designer with the possibility to change the drinking container almost infinitely without much effort. In this way, one is able to address the dilemma presented in Chapter 1, illustrating an example of elongated design.

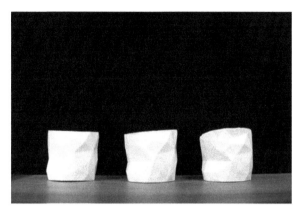

Figure 5.5 Three unique cups. The irregular faceted cylindrical form is a 3D-printed clay cup. Looking at the form we find a deformed cylinder where the designer has pushed and pulled a low-resolution file to achieve uniqueness in each and every cup. Design: William Kempton; photo: Inger Steinnes.

5.6.4 Intersectional Forms

Intersectional forms overlap or cut into each other; they offer a way of generating combined properties from two or more geometrical forms. The task given to students was to model one round and one edgy form and to orient them in space for the best intersectional form (Fig. 5.6).

Beginning with two different forms, it's startling what you can get out of combining them in an interesting way. Students had to be sure the forms cut into each other. The common mass in this we could call an intersectional form or core. Properties of the "core" or intersectional form are given by the orientation of the first two forms. Composing and modeling it manually in clay is a key technique to train form analytical skills. But how does it function when transposed into the digital fabrication tools and design actions entailed in taking up AM?

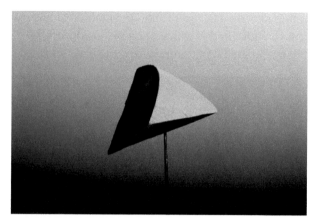

Figure 5.6 A "core" from a tetrahedron and an extensional ellipsoid common mass in an intersectional form. Photo: Maria Karlsen.

From teaching experience, it is clear that freeform fabrication enabled via digital software and AM print technologies opens this field even further, with the possibility of iterating fast and with successive and reshaped physical manifestations. We see that this form task is especially suitable for multiple digital versions. Intersectional forms have complex parting lines as their main features. Some volumes with multiple parting lines would have resulted in the use of extreme production tools with several parts at a considerably greater expense than with AM.

A well-known expressive form directed from the established production technology is self-explanatory, comforting, and understandable. "Attractive design comprises a thoughtful balance between novelty and typicality," observed Schifferstein and Hekkert [12]. Yet, as design educators and researchers, we need to face a paradox: the fact that users like typicality (something familiar) in what they buy is somehow contradictory to the drive for novelty. Innovation does not always give users what they want. The freedom of form through the use of the EoF model imposes a risk of making new expression that is alienating for the users who are present. The famous MAYA principal, proposed by Raymond Loewy in 1954, describes the balance. MAYA is an acronym for "most advanced, yet acceptable." It is precisely this balance that we ask our students to learn to perform and we extend this into PD realized through AM (Fig. 5.7).

Figure 5.7 Thirty examples from an EoF workshop. Photo: Maria Karlsen.

An interesting thought would be how we as users will react when new, more provocative shapes will emerge. Will the balance stated in the MAYA principle be challenged? Will there be an increased acceptance for a faster change in forms than previously? 3D printing certainly opens up for this, and for some types of products, for instance, those generated by computer algorithms (see Fig. 3.21 in Chapter 3) it would be a prerequisite that people would accept new forms emerging at a much higher pace than before. This leads back to one of the questions raised in Chapter 1, are we ready for these new types of fast-changing products? Is it a good thing that the life span of a product will be even shorter than today, and is that a sustainable approach?

5.7 Familiarity

Already in *Vision in Motion*, Moholy-Nagy's book from 1945, shaping sensibility was realized by making small "volume families" [16]. The samples in Fig. 5.8 can be described by operations from Akner-Koler's model but are not one of the seven form pairs in the EoF model.

Figure 5.8 Form families with similar overall proportions and properties but diverging details, cuts, and textures. There are also common features as resemblance to animals or more morbid parts of the body within the groups. First-year student's work, 2016. Photo: Maria Karlsen.

Features and properties link a number of objects together. The metaphor "form family" explains the phenomenon familiar features are in common. For these artifacts, whether it is the way they are distorted, their silhouette, or some important detail, we perceive them as related.

Relationships here are meant as different features working in symbiosis. Shapes, parts, texture, cuts, resemblance, or familiar imprints help produce understanding of belonging. Industrial designers often work with new or established brands. The physical objects represent the brand as well as all touch points linked to it [10]. We acknowledge the wide span of formal possibilities linking objects together and teach students to value their 3D work beyond sticking logos onto ready-made objects.

AM plays an important role in such processes and their artifact generation through the possibility that lies in programming form "elasticity." Minor changes to every print provide uniqueness but also maintain identity throughout the series. Having found few others venturing into what would seem an obvious domain for PD, we hope that this chapter has begun to point in some fruitful directions for centering on the aesthetic and form as regards AM. We have sought to show how relations between PD and AM may be more fully taken up through attention to the aesthetic and 3D visual form in particular. We have suggested ways that this may be understood: we raised some of the benefits of connecting models such as EoF to the "additive" and generative potential in AM tools and technologies.

Our exploration so far concerning connecting 3D visual form finding and AM points to a number of possible directions for further consideration: produce, create, and perceive. First, there is clearly room to further investigate the possibility to produce fully free forms using AM. Second, one might venture into exploring how to enact radical ways to create such forms. Third, there is space to examine how such fast and radical form development may be perceived.

We have not yet taken up the many of the possible routes of inquiry in seeing how the model of 3D form giving mentioned above might itself be read more closely and transformationally via 3D software. There is clearly work to be done in applying such a model and exploring further the emergent aesthetics of toggling between the physical and the digital, the conceptual and the iterative. These have implications for form finding and form giving and their wider connection to PD. This naturally leads us over to the next chapter, "Potential of Additive Manufactured Products in Building Brands."

References

1. Akner-Koler, C. (1994). Three-dimensional visual analysis. Department of Industrial Design, University College of Arts, Crafts and Design, Konstfack, Stockholm.

2. Akner-Koler, C. (2007). Form and formlessness (Doctoral thesis). Axel Books/Chalmers Technical University, Göteborg.

3. Akner-Koler, C. (2012). Expanding the boundaries of form theory and practice. In Steinø, N., and Özkar, M. *Shaping Design Teaching*. Aalborg: Aalborg University Press, pp. 129–145.

4. Arnheim, R. (1974). *Art and Visual Perception: A Psychology of the Creative Eye*. Berkeley: University of Califronia Press.

5. Capjon, J. (2004). Trial and error based innovation (PhD thesis). AHO, Oslo.

6. Ching, F. (1979). *Architecture Form Space and Order*. New York: Van Norstrand Reinhold.

7. Gail Greet, H. (2002). *Elements of Design*. New York: Princeton Architectural Press.

8. Garden, J. (2016). Additive manufacturing technologies: state of the art and trends. *International Journal of Production Research*, **54**(10), 3118–3132.

9. Hansen, F., and Priska, F. (2016). 3D printing as a ceramic craft tool in its own right. In Mäkelä, M. (ed.). *Ceramics and its Dimensions: Shaping the Future*. Helsinki: Aalto University School of Arts, Design and Architecture, Vol. 4, pp. 114–128.

10. Hestad, M. (2013). *Branding and Product Design: An Integrated Perspective*. London: Gower Applied Research.

11. Jones, M., Seppi, K., and Olsen, D. (2016). What you sculpt is what you get: modeling physical interactive devices with clay and 3D printed widgets. In *Proceedings of the 2016 CHI Conference on Human Factors in Computing Systems (CHI '16)*. ACM, New York, NY, USA, pp. 876–886.

12. Schifferstein, H. N. J., and Hekkert, P. (2008). *Product Experience*. Oxford: Elsevier.

13. Schneiderman, B. (1986). *Designing the User Interface: Strategies for Effective Human–Computer Interaction* (1st ed.). New York: Addison-Wesley.

14. McCullough, M. (1996). *Abstracting Craft: The Practiced Digital Hand.* Cambridge, MA: The MIT Press.

15. Michl, J. (date). Form follows WHAT? The modernist notion of function as a carte blanche. Available at http://janmichl.com/english-only.htm (accessed on April 16, 1925).

16. Moholy-Nagy, L. (1945). *Vision in Motion.* Chicago: Hilliston & Etten Company.

17. Steinø, N., and Özkar, M. (2012). *Shaping Design Teaching.* Aalborg: Aalborg University Press.

18. Papadakis, A., Cooke C., and Benjamin, A. (1989). *Deconstruction: Omnibus Volume.* London: Academy Editions.

19. Torres, C., Campbell, T., Kumar, N., and Paulos, E. (2015). HapticPrint: designing feel aesthetics for 3D printing. Presented at *UIST 2015*, November 8–11, 2015, Charlotte, NC, USA.

20. Wong, W. (1993). *Principles of Form and Design.* New York: Van Norstrand Reinhold.

Chapter 6

Potential of Additive-Manufactured Products in Building Brands

Monika Hestad[a] and Viktor Hiort af Ornäs[b] †
[a]*Brand Valley AS, Oslo, Norway*
[b]*Chalmers University of Tecnhology, Gothenburg, Sweden*
monika@brandvalley.no

My first attempt to use additive manufacturing (AM) in a brand building and product design context was when I was an industrial design student.[1] This was in 2004, and the technology has, as we see in other chapters in this book, come a long way since then. I was collaborating with a porcelain manufacturer and had decided to use this technology to get a prototype of the plates I designed. In particular, I wanted a tool that allowed me full form freedom and precision in developing the form in the design.

The aim of the project was to use inspiration from the Norwegian fish industry to develop a new brand concept and a ceramic that communicated the essence of the brand developed. Core values of the

---

†We sadly lost Viktor to cancer on July 24, 2016. He was a thoroughly committed researcher, a great intellectual, and a very good friend, who contributed to the very end. His ideas will live on.

[1]When asked to reflect upon the role of additive manufacturing in relation to branding, one of the authors went back to her time in the university.

Additive Manufacturing: Design, Methods, and Processes
Edited by Steinar Killi
Copyright © 2017 Pan Stanford Publishing Pte. Ltd.
ISBN 978-981-4774-16-1 (Hardcover), 978-1-315-19658-9 (eBook)
www.panstanford.com

brand concept were defined as courage, contrasts, dreaming, *and* sensation. *So, how do these value words translate into a form? To make a long story short I found the expression I sought after when experimenting with paper. Crunched paper has interesting lines, and combining these in a way that gave a feeling that a plate was broken created a novel aesthetic. This had both a complexity and a simplicity at the same time. I worked first in clay and ended up with a form that was asymmetric, a combination of concave shapes and strong accents where the shapes met, a form that in theory should be perfect for the* full form freedom *that the technology offers.*

As many young designers, I found it challenging to transfer what was expressed in clay, as my skills of how to use the technology were not yet developed. I was working in Rhino Zero, and the surface-based tool made it possible to build the shape. However, this program also has a weakness as complex shapes often get holes *in the surface. The complexity in the form with a variety of concave surfaces was a nightmare to build. Many days and nights later I managed to get something that was possible to print. The shape was close to the intended form but still not what I had in mind. It did not communicate what I set as a goal for the brand concept. After another week in the workshop, which involved a massive amount of filling and hours of polishing the form, it finally gave the impression I wanted (see Fig. 6.1). Not exactly rapid prototyping!*

Figure 6.1   Final result of additive-manufactured prototypes of ceramics.

The anecdote gives a familiar scenario in a learning context. When building a product for a new brand concept, the designer has a variety of decisions to make. Intention, desired aesthetic, and ambition can to a certain level be controlled, but other factors also influence the final result. In this case it was a combination of a lack of skills, how far the technology had come, and how much time the designer had available. In a company context there will be even more drivers behind the product to be aware of. There are external drivers, such as the competing situation and trends in the market. There are also internal drivers, such as what the company has produced before [2] and what kind of manufacturing equipment it has available.

The promise of additive manufacturing (AM) is that it introduces new material qualities, production techniques, and ways of interacting with the customer. It offers form freedom and opportunity to produce on a small scale and shortens the distance from when the product is designed and manufactured [16]. From a brand building point of view, there are several opportunities; AM will allow the designers to design a product that is in line with the desired communication of the brand. It gives the design team a possibility to express the identity of the brand in the form and to make something that is distinct to one brand identity in particular. It will help the team to create resemblance across the product portfolio. Small-scale production could enable production of limited editions and allow items that are custom made and where the consumer can affect customizations. AM could also contribute to another layer of complexity in how the product is being designed as well as being manufactured. Within manufacturing AM is still relatively new and offers a different set up to conventional manufacturing processes [17].

What real benefits would be in building the brand is yet to be explored. Many of the companies the authors have researched and worked with have still not developed a sophisticated system of identifying how the product should contribute in building a brand. So to ask how manufacturing of a product could contribute to building a brand may be an even further stretch.

In this chapter we will first give a short introduction of basic brand theory. It will offer two frameworks that could help the company to understand its own products in relation to the brand. As the anecdote shows AM is only one of many drivers affecting the product design. Other factors, such as attitude, ambition, and know-how, may have a bigger influence on what is designed and how this design contributes building the brand. The frameworks will help to analyze these contexts. These frameworks will be used to analyze two cases—Mykita and pq by Ron Arad—that are using AM in producing eyewear, and analyze what this adds to their brands. The chapter concludes with a brief discussion on the potential from a brand building perspective to use AM in developing products.

6.1 Product Role in Brand Building

In this first section of the chapter we will give a brief overview of what a brand is, the role of design elements in building a brand, and the product's role in building a brand.

6.1.1 Role of Design Elements in Building a Brand

According to the American Marketing Association's much quoted definition a brand is a "name, term, design, symbol, or any other feature that identifies one seller's good or service as distinct from those of other sellers" [3]. In this definition a brand is defined as all those things that make a customer recognize and associate the goods or the services with one company instead of another. It is something that helps a company to distinguish its goods or services and make them unique. A brand as such is intangible as it resides in the consumer's mind, but to trigger the idea about the brand, it also needs to be made concrete through a feature, such as a design element. Toni-Matti Karjalainen and Anders Warrell found that design elements can both give direct identification and lead to an association [5].

Using a well-known and iconic brand such as Coca-Cola, it is easy to identify multiple design elements: the red and white colors on a dark-brown background, the Coca-Cola typography, the silhouette of the bottle, the texture of the glass bottle, etc. These distinguish Coca-Cola's goods from Pepsi's and makes Coca-Cola distinct and thereby easy to recognize. By using the same design elements coherently over time the company manages to take ownership to these elements. The success of making a brand is when the customer recognizes the design elements and these trigger associations.

For the design elements to become vehicles for brand meaning, they need to be associated with a core message or associations that the company would like to evoke in the customer. *Happiness* has been one of the core values the Coca-Cola Company would like the Coca-Cola brand to be associated with. Looking through various commercials the Coca-Cola Company is retelling the story of the Coca-Cola brand in various ways but always in a cheerful and happy tune. In this way, the company manages over time to establish an emotional connection with the customer around happiness.

6.1.2   Brand Story and Product Story

Telling stories that involve the brand is key in brand building [14]. The researcher Douglas Holt (p. 3) [14] lists four different types of authors in brand building: the company behind the brand, culture industries, intermediaries (critics and salespersons), and customers. A common trajectory when building a brand is that the company starts with telling a story about how the product was invented, or a founder story. This is retold by others interacting with the brand as retail persons when selling the product to a customer. Hopefully, as a result the consumer is buying into the vision of the brand and is sharing this story with other consumers.

Another trajectory where the customer takes a lead is, for example, by the shoe brand Dr. Martens. On its website today you could read that Dr. Martens is about "people who possess a proud sense of self-expression. People who are different" [11]. This is not what it started with. The shoes were first popular among housewives in Germany. To make a long story short, after a redesign and relaunch in the U.K. in the 1960s, they were picked up by a cultural movement that showed pride in being the working class with a rebellious attitude. In the next decades the shoe started to be popular among several underground movements (e.g., punk and glam) and by this became a cultural symbol. In this case, it was not the company that created a story and sold it to the youth, but customers with a strong identity and self-awareness that themselves picked the shoe to represent themselves. Today this is the story that Dr. Martens tells its customers.

Dr. Martens and Coca-Cola, used as examples so far, represent two different brands. What is similar between these two is that they could both be seen as cultural icons and that the brand is something beyond the product itself. The glass bottle in Coca-Cola and the Dr. Marten boots could both be seen as 3D symbols representing the brands, and seeing these might trigger the association of the brand. In both these brands the brand story is on a higher abstraction level than the product story. The product story about the Dr. Martens shoe could be "a robust and sturdy boot, with a comfortable sole made for hard manual labor," while the brand story might be "people that are different and rebellious." However, personality traits that could be used to describe the Dr. Martens boots, such as *robust* and *tough*, resemble well the core values of the brand.

In other brands that are about the use of the product or how the product is made, there is a closer resemblance with what the brand story and the product story is about [15]. The Finnish brand Fiskars is well known for its ergonomic tools, such as its scissors with the orange handles. The product story is about ergonomic use, and the brand story is about an ergonomic and simpler life because of these tools. In this case the brand becomes less abstract than in the case of the above examples of Dr. Martens and Coca-Cola. In between there are also those brands that are not directly about the products but related to the products, like being about the origin, heritage, or people involved. The American razor brand Gillette plays on its consumers and what the product means for them, exemplified by the slogan "The best a man can get."

The brand and product relations can be simplified to three core categories (see Fig. 6.2). The first level is where the product story is the brand story. These are often companies that are driven by innovation or technology, where Fiskars is an example of this. The second level of abstraction is when it is in the wider context of the product, which Gillette could be an example of. The third level is when the product plays a symbolic role and the brand is cultural or driven by a myth. Both Dr. Martens and Coca-Cola belong in this sphere. The reality is seldom as clear-cut as a model will be. All of the brands mentioned also have elements of the other stories in their branded stories. For example, Dr. Martens still tells the story about the functional and comfortable sole that was invented; it tells the story about the inventor in its name, Dr. Martens, as well as plays on its strong heritage as being the chosen one among the rebellious over decades.

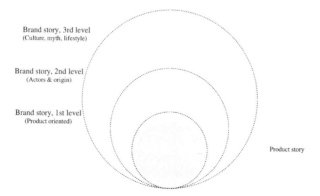

Brand story, 3rd level
(Culture, myth, lifestyle)

Brand story, 2nd level
(Actors & origin)

Brand story, 1st level
(Product oriented)

Product story

Figure 6.2 Brand–product abstraction model.

6.2 Additive Manufacturing as One of Many Other Drivers That Affect a Product's Role in Building Brands

The anecdote at the beginning of this chapter presented how a design student experienced that other factors than the will to communicate brand values in form affected the final design. In a company context even more factors than own ambition and skills will interfere with the design. In addition, there is, as mentioned before, complexity in designing a message in the form and duality in what the intended message and the actual message are. In this part of the chapter a model that seeks to explain this complexity will be introduced.

6.2.1 Actual and Intended User Experience

In Section 6.1.1 the role of design elements was briefly described. These play an important role in building the brand as this is were part of the story can be made concrete and this will trigger the associations to the brand. These design elements could be the product as a whole, as seen with Dr. Martens, or elements such as the sole or the yellow thread. They could also be how elements are combined together, proportions or repetitions occurring in a certain way. They can also be less explicit and be more a feeling or an association that the composition of the elements triggers or if the brand is of certain personality traits that are expressed in the form. Design elements in a product are also more than visual cues. They could be a tune, a taste, a tactile experience, or a smell. Further, how customers interact with the product can also serve as a reference to the brand.

What the brand represents and what these design elements communicate may be interpreted differently from user to user. For one user, Coca-Cola could be childhood memories of sharing a Coca-Cola with his or her grandparents. For another it might be the feeling of happiness when seeing the commercial of the Coca-Cola in which Santa Claus promises us another family Christmas. The brand as such becomes the experience the user has had, as well as will have, when engaging with representations of that brand.

A product and its design elements seen from the perspective of a user are also more than its physical properties. The product allows people that are using this to express values, an identity, or a shared lifestyle that is associated with that brand. It offers certain benefits, allowing people to do or feel something. Users will inevitably experience things differently, but by designing products that appeal to the senses, which engage their users in interaction and carry certain symbolism, companies can scaffold for user experience. In building a brand, the company needs to know which user experience it is planning for and will fulfill, as it is ultimately these that will be the brand.[2] The user experience the design team has in mind might or might not be what the actual user experience becomes (see Fig. 6.3).

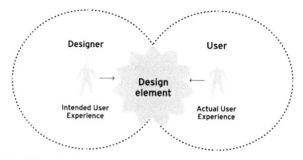

Figure 6.3 What the designer would like the design to represent is not always what the user experiences.

---

[2]To help us to understand the user experience, Jordan [13] introduces a four pleasures analysis of products: physio-pleasure, socio-pleasure, psycho-pleasure, and ideo-pleasure. These are based on Lionel Tiger's [27] framework of four pleasures. These four categories are useful in understanding the role of the product in building the brand.

Physio-pleasures are bodily pleasures from senses (touch, smell, taste, vision, and sound). Examples are removing discomfort with a product and aiming for an aromatic smell or tactile experience when touching an object.

Socio-pleasures are pleasures that arise when engaging with other people or society. These are about belonging to a certain group or being accepted. For example, new technology makes it easy to communicate and stay in touch with a large group of people.

Psycho-pleasures are pleasures that are about cognitive or emotional response. A product can create discomfort, stress, or annoyance. For example, when the national rail is always delayed or when replacing the train for a bus starts to be the standard instead of the expected, this negativity for the company starts to become associations that are triggered when the brand is mentioned. (*Contd. on next page*)

6.2.2   Internal Drivers

There are several drivers internally that ultimately shape the product (Fig. 6.4). When designing a product, multiple considerations and limitations will shape what it becomes. From where these emerge depends on the context the product development process is part of. Internal constraints may come from production method, material, or internal know-how. Others may stem from the organization, who the decision makers are, work procedures, information flow, and organizational structures, budget, or culture of how to do things. Other central drivers are whom and what the designers draw inspiration from, their ambitions for the product such as the price point, and the position they seek in the market.

Manufacturing will often set limitations and standards in what is possible to design. However, this is one of the promises with AM that many of the limitations that come with conventional production techniques may change. Designing for ease of manufacturing is important, and manufacturing techniques bring about alterations on the product [16]. Further, several alterations could happen in the implementation stage to meet certain budget requirements or to simplify the production process.

Another promise that comes with AM is that this will make the design process more hassle-free as the design and manufacturing are a closer fit. An example of how manufacturing constraints restrict the overall form freedom of the designer is a design of a bottle. The angle that will define the height and form of the "shoulders" is defined by production limitations. This could affect how "proud" or "strong" a bottle will appear. Constraints like this may change the expression of the bottle and affect how unique or generic the bottle will, at the end, appear.

In brand building the heritage, what has been done in previous designs, and other activities play a major role. It is important to build a coherent style across a product portfolio. This relates to the kind of products that are developed, the concept behind the products, the dynamics of the change, and how the products are part

Ideo-pleasures are pleasures that are about people's values. They can tell people what is important for them, what their tastes are, and what moral standards or personal aspirations they hold. Any fashion brand will offer an identity to people that could help them express their tastes, likes, as well as values.

of building the brand. Over time the company may have identified several traits and references that it sees as important for its brand. How the company relates to them and the importance of them in the marketplace become important drivers in the design process. In an organization, how things used to be often informs how decisions are made and what will be done in the future.

6.2.3 External Drivers

The company and the design team will also be affected by several external drivers and constraints, such as trends in the design community, trends in the product category, as well as trends in society (Fig. 6.4).

When a product is designed it will be categorized with similar products and these together form a product category. Coca-Cola is part of the beverage category but can also be seen as part of the soft drink category. In this category it is a global market leader and has a strong influence on how the category changes. Another product that would like to be part of the soft drink category may try to seek to fit these so-called unwritten category rules.

Whether the category rules exist may be a myth, but some resemblances exist across the category. It can be the use of bright colors, how the name is written, and how much of the label should show the name. It can also be about the product specifications of where they can be found on the label, the dimensions of the bottle, how the customer should use it and hold it, and so on. All of these can give the consumer an idea about it belonging to a certain category and can be called *category references*. When the category changes, for example, by using a different material, or when every beverage gets its own distinct profile bottle, the consumers' understanding of the product as part of the category will also change.

Another major external driver is the intended users or customers: their values, behaviors, and identity. The choices they make and how they interact with each other, with other companies, or with the brand will also influence the design process.

Major political, environmental, social, technological, environmental, or legal changes could influence the opportunities in the marketplace as well as the company. A technology such as AM will have the potential of affecting the company and the design process both directly and indirectly—directly if the company makes use of this and indirectly if its competitors start to make use of this—and changing customers' expectations of the product category. This could potentially lead to a disruptive change where the company if it does not change quickly enough may find itself obsolete in the market space [7].

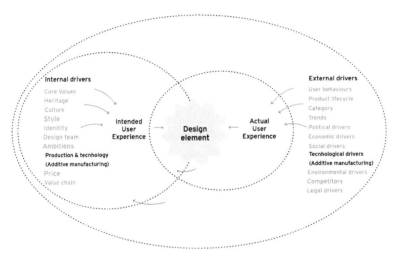

Figure 6.4 DUD framework: Design elements, User experience, and key Drivers. AM is one of many other factors influencing the design and interpretation of a product.

6.3 How Additive Manufacturing Is Used in Building a Brand

At the Oslo School of Architecture and Design several experiments in what AM could offer the product design process have been made through master and PhD work. A project by Kathinka Bryn Bene explored the possibility of how AM could be utilized in design of eyewear with the purpose of adding value to the product [5]. In this experiment, design customization was used as the main point for exploration. One of the main challenges was to balance the freedom for the users to choose a design according to ergonomics or taste, at the same time controlling the quality of what was designed. The eyewear was designed in a way that it could balance the identity of the user and the brand. Bene chose the bridge (the element between the lenses) to create a y-shaped detail as an intended brand signature. In addition, she also chose a distinct atypical color for her sunglasses. According to Killi [5], these "two features together create a strong recognition effect and allow an elasticity of the form." As such, the identity of the brand is maintained while the form of the frames could still be adjusted (see Fig. 6.5).

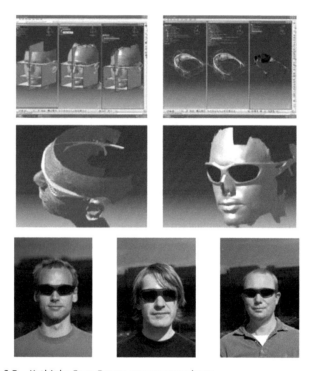

Figure 6.5 Kathinka Bryn Benes eyewear products.

The study by Bene shows the opportunities that lie within using AM consciously in building a brand. Eyewear in general is interesting as a case as this is an area where the product is a lot about the identity of the user. The user is often using design or certain brands to communicate his or her style. Today, a decade after Bene's experiment, there are examples in the eyewear category that are using the benefit offered by AM. These include the Berlin brand Mykita and the Ron Arad pq eyewear. In the following sections, the frameworks introduced earlier in the chapter will be used to investigate the role of using AM in building these two brands. The material that this study utilizes is mainly from open sources such as visiting the stores, printed marketing material, the glasses themselves, and what is available online.

6.3.1 Mykita

The first brand that will be examined is the Berlin-based eyewear brand Mykita, which was founded in 2004 (Fig. 6.6). Today, it has a global reach with flagship stores in cities such as Berlin, New York, Monterrey, Vienna, and Tokyo. In addition, it is sold through multiple stores across the globe. In 2004, it launched its first metal-frame collection, Collection No1. Later it launched several more collections and in 2010 introduced its first AM eyewear, named Mykita Mylon (see Fig. 6.7). In 2016, it took the benefits offered by the technology one step further and introduced the service My Very Own: "MY VERY OWN integrates three digital technologies—3D scanning, parametric design, and additive manufacturing—to tailor the design and fit of a pair of glasses to the individual topography of each face" [23]. My Very Own is created in partnership with Volumental, which is described as the "leader in the field of product customization using 3D scans of the human body" [23]. In the examination, we will use the DUD framework and examine design elements, user experience, and key drivers (Fig. 6.8).

Figure 6.6 Mykita. Photo: Mykita.

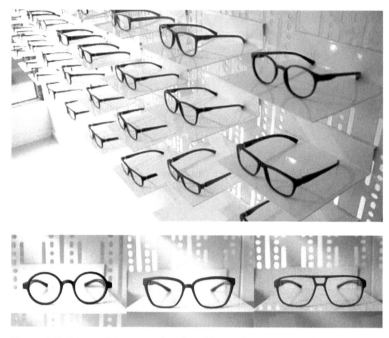

Figure 6.7 Three different pairs of Mykita Mylon, Ses Optikk, Oslo. Photo: William Kempton.

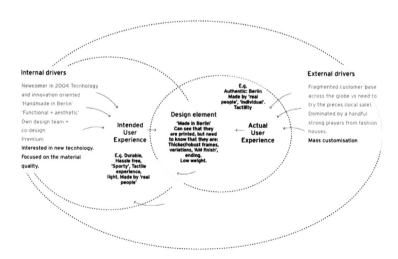

Figure 6.8 DUD analysis of Mykita Mylon.

6.3.1.1 How AM is used in the products

6.3.1.2 Design elements

For people who are not familiar with AM it might be difficult to see that the Mykita Mylon eyewear is not made by conventional manufacturing. The product is produced by selective laser sintering (SLS) (described in Chapter 2). The glasses are light, as promised in the marketing material. On a closer look, the texture reveals that it is made by laser sintering as well as through a clear cut on the handles of the glasses (see Figs. 6.9 and 6.10). The Mykita brand is a fashion-driven eyewear brand, and the glasses follow the latest trends in the market along with the latest in new technology [9, 18]. Its reasoning for using this technology is, "With this technology you can create a 3D product exactly in the shape you want to have it, without additional tooling costs. It gives us the possibility to create a product that from a performance perspective is better than anything out there when it comes to sports eyewear" [8].

Figure 6.9 Mykita Mylon, details, Ses Optikk, Oslo. Photo: William Kempton.

6.3.1.3 User experience

Highly prominent in Mykita's marketing material is the fashion association. The models wear elegant clothes and ride elegant bikes to indicate the lifestyle that follows the fashion brand. It has strong competitors in major fashion houses, and this also influences how it has chosen to present its own products. Mykita plays on the functional benefits that the material gives. In its own words, it is "exceptionally light, durable and adaptable," which is relevant in sport activity. The reason for using AM seems to be mainly the functional benefit with the material, which again will give the users a product that has low weight and is durable. The combination of description of the technology used and the fashion "look and feel" are coherent across its marketing material.

Figure 6.10   Details on Mykita Mylon. Photo: Mykita.

Since the product is made with new technology it may also offer the user an opportunity to show that it is forward oriented; the "Handmade in Berlin" tag gives the brand a certain authenticity.

6.3.1.4   Drivers

The company was founded in 2004, and since then it has launched several sub-brands on the market. Exploring new technology and craftsmanship seems to be the essence in what drives the product design process (see Fig. 6.11). The foundation of the first glasses that it built was No1, which has a technique of folding that is similar to origami. This gives the glasses a unique look and makes the users' interaction with them slightly different. Their keen interest in innovation and exploration in new technology is an important driver in new development. This seems to be in line with what is important for the market as well, as presented by Forbes author Joseph DeAcetis: "Forecasting the future of eyewear is not only about product familiarity. More ingenuity and more technology seem to be the common traded views to reach modern consumer demand" [9].

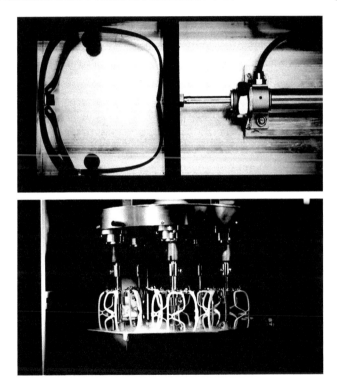

Figure 6.11   Manufacturing of Mykita Mylon. Photo: Mykita.

The company is guided by modernist design principles, using modern manufacturing techniques merged with a strong interdisciplinary thinking. "We also look at different industries, materials and other technological fields when we work on new constructions or new hinges. This modern approach then meets with traditional, precise classic craftsmanship" [8].

Diversity is important for the company, and the development of the business through a vertical strategy is highlighted as a strength as this gives it an opportunity to control the quality in every part [9]. It has gathered know-how and expertise under one roof in the Mykita Haus in Berlin. This allows the company to experiment internally but also to invite collaborators to design together with it [8]. As part of its strategy it has actively sought collaboration. It has worked with the German fashion designer Bernhard Wilhelm and the Paris-based fashion house Maison Margiela's creative director John Galliano, to mention just two.

The importance of exploring new technology may also be the reason why it has started with AM. The promise presented in the campaign is that it seeks "bridging the gap between fashion and sports." "Handmade in Berlin" seems to be a stronger added value than AM.

Exploring new technology and collaboration gives Mykita an opportunity to learn what AM can offer beyond the material quality utilized in Mykita Mylon. In 2016, Mykita launched My Very Own. With this development, it got an opportunity to customize glasses to make the perfect fit. My Very Own draws on parametric modeling, which allows for variation with certain dimensions driving the design. From a brand point of view, it becomes central to identifying signature elements and proportions and interpret brand intent into "what can we allow to vary" and "what relations need to be controlled."

6.3.1.5   The Mykita brand story

The first impression when visiting the website, the shop, and the display is that this is a fashion-oriented brand. The shops are described as art galleries, and the social media profile shows various celebrities using the glasses. The campaign in 2016 was named "sequences of island portray a lifestyle." The lifestyle that is portrayed is sophisticated, yet casual and informal. The clothes the models wear replicate the simple cuts and lines from the eyewear. The brand story is not only about the products but also about the lifestyle that can be experienced and how the products enable this.

The first office of the company, which was in a former day nursery, gave the name Mykita [9]. It is a play on the words *My* and *Kita*, with

Kita being a common abbreviation in German for a *Kindertagesstätte* (nursery).

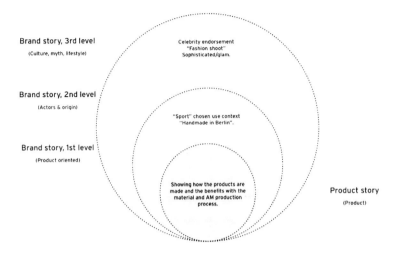

Figure 6.12 The Mykita brand plays on all three levels.

An important signifier of the brand is "Handmade in Berlin." According to one of the founders, Moritz Krüger, being established in Berlin has influenced the brand's philosophy: "Berlin is constantly moving and changing. It is a lot about imperfection and the permanent process of reinvention and transformation, making the best out of limited resources . . . So, the philosophy to produce everything in-house arose out of necessity" [8]. It seems as though the brand would like to be portrayed as a sophisticated craftsmanship brand based in Berlin.

Going one step deeper and reading about Mykita, the exploration into diverse technology and how the eyewear is made play a central role. On the social media channel YouTube it is presented that "MYKITA is a modern manufactory that combines precision craftsmanship with new technologies." The company is sharing how the frames are made and how the team is working tighter and highlights the importance of the "holistic business models" that are about gathering different expertise under one roof [20]. This one roof is in the case of Mykita not only a metaphor but also an actual building referred to as Mykita Haus.

The response from people engaging with material from Mykita is finally that there is a brand that is "handmade in Berlin" by "real" people contrary to machines. The manufacturing of the products and that the company is showing the people behind it makes this brand unique compared to others.

Looking at the brand through the brand–product abstraction models, it plays on all levels (Fig. 6.12). Mykita seems to be aiming at a story that combines the lifestyle of young fashion-oriented people with the interest in well-crafted and designed products. In the presentation material the majority of the text is about how the products are made and what kind of technology is used when making the products. The products and how these are made are central in telling the story about the brand. AM is not that prominent in the overall brand story, but the use of new technology is and being driven by the principle that "the technical solution must also be an aesthetic one" [22]. Their combination of distinctive design and novel engineering has won them interest from fashion's most experimental designers and brands [24].

6.3.2 pq Eyewear by Ron Arad

The second brand examined in this chapter is the pq eyewear by Ron Arad. In 2013 Assaf Raviv, an Israeli eyewear manufacturer, managed to convince designer and architect Ron Arad to create a line of eyewear.

Ron Arad is a well-known figure in the industrial design scene (see Fig. 6.13). He has headed the design products department at the Royal College of Art. His work has been described as rough and scary, which applies as much to his creations as his personality as he is said to be a "big, gruff, bovver-booted sort of bloke" [12]. One of his most famous pieces is the bookworm bookshelf produced by Kartell.

Ron Arad was initially hesitant as he thought the product might force him into a culture of fashion [6]. However, Ron Arad got the freedom needed to challenge status quo, and today the pq eyewear presents itself with the tagline "anti-ordinary" (see Fig. 6.14). pq today has three product lines: the highly adjustable A-frame and two lines (B-frames and D-frames).

Figure 6.13   Ron Arad. Credit: pq by Ron Arad.

Figure 6.14   pq by Ron Arad, website. Credit: pq by Ron Arad.

All of these are based on AM. pq's first product, the A-frame, was complex in that several parts had to be combined to allow for adjustment. The adjustment is made possible by an A-shaped connection, which also serves to visually differentiate the product from other sunglasses. B-frames and D-frames, instead, draw on AM. This allows for fewer parts. An important part of the making of the eyewear is how it interacts with the customer through a process they call 3 Ps: perfecting, personalizing, and playing. This process is facilitated by the designer by providing a platform from which customers can create their own glasses. On this platform they can add their personal measurements and choose between predefined styles and colors.

In the following section we will use the DUD framework and examine Design elements, User experience, and key Drivers (Fig. 6.15).

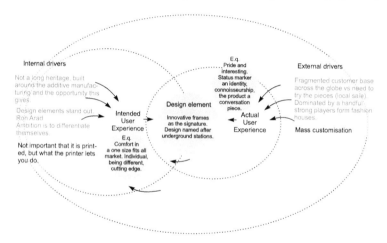

Figure 6.15   DUD analysis of pq by Ron Arad.

6.3.2.1   How AM is used in the products and in branding

6.3.2.2   Design elements

The different pq product ranges use different design elements to stand out embedded in the frames in the three different glasses that the company has manufactured.

A-frames (see Fig. 6.16) are signified by an A-shaped element between the screens, allowing the frames to be adjusted to fit any face by sliding the lenses along an A-shaped wire over the nose. The same wire forms a hinge mechanism that automatically closes the arms when the glasses are taken off.

Figure 6.16 A-frame. Credit: pq by Ron Arad.

B-frames are made by two composite, visually explicit lines (see Fig. 6.17). Expanding on the possibilities of the SLS production technology, the TwoLiners are made from top and bottom lines. These are connected at the temples and separated in the front, holding each lens.

Figure 6.17 B-frame. Credit: pq by Ron Arad.

D-frames (see Fig. 6.18), launched in 2015, are made as one piece, made possible by AM: "There are no hinges, no pins, and they are lightweight with a total flexibility, made in one piece" [26].

Figure 6.18 D-frame. Credit: pq by Ron Arad.

For some of the products there are additional themes, for example, the naming of A-frame models is based on London subway stations.

The brand pq is an important design element (see Fig. 6.19). pq presents an interesting story around the name: "Nestled next to each other in the middle of a hostile alphabet, the mirror of each other, meaning nothing, and importantly meaning nothing in Japanese. pq stands for nothing but the shape of a pair glasses" [10].

Figure 6.19   Details. Credit: pq by Ron Arad.

6.3.2.3   User experience

The brand promise is that the user will feel special by having his or her own individually fitted eyewear, which tells the story about a designer known for his cutting-edge style. The marketing material plays on individuality and the importance of being different.

pq's own description of its users is as follows: "Beautiful, intelligent, generous, funny, charming, extraordinary, special. I'm thinking about positive words. Educated, cultured, progressive, liberal, advanced. Did I say also generous? Yes, of course" [1]. This means that the brand is targeting people who want to be different and are willing to pay the extra for a pair of glasses that signals their identity.

AM allows the company to offer a user experience that is unique as the users experience themselves as part of designing their own glasses. This also provides them perfect custom-fit glasses in a one-fits-all market. The made-to-measure opportunity promises superior comfort.

The glasses are a functional item, but they are also a piece of fashion, which plays on identity, and a status marker. By choosing a brand that is not from one of the big fashion houses, it signifies a certain level of connoisseurship. The use of nonstandard structural signature elements such as the A- and spine joints has the potential of fascinating and capturing interest.

6.3.2.4   Drivers

Internally, pq does not have a long tradition to relate to, and AM has been essential in setting up the company. This means it can allow itself to explore the potential AM gives. The form freedom of AM allows production of some details, such as the hinge on the B- and D-frames, which would be difficult to produce with other techniques. AM is also important as it allows pq to produce in smaller batches and a wide range of variations, which tooling costs may have prevented with other manufacturing techniques. This may allow a wider range of variation within each product line, supporting the message about "being different" but also about a better fit through making it possible to tailor sizes to the customers' heads.

pq builds on the 3 Ps, which will be an important part of the user's own experience of the brand. The perfecting is the algorithm that makes it possible to make custom-fit glasses. The personalization is about how the customers could make their own glasses by scanning their heads. The playing is about how the customers together with pq find the perfect playful match.

Externally there are several influences affecting pq. pq acts in an industry that has been dominated by a handful of players of big fashion houses. The requirements and design space for eyewear are rather well explored. pq's products face similar requirements as other eyewear, which will affect the design of the eyewear.

pq wants to differentiate itself from the competition. Building on Ron Arad's reputation this means to be bold in its design options. Ron Arad has a history of working with products that have a technical appearance, such as making use of light-emitting diodes (LEDs). What the competitors do influences therefore less directly how they behave, but it will influence them indirectly by trying to be different from what else is out there (see Fig. 6.20).

Figure 6.20   A form that differentiates. Credit: pq by Ron Arad.

6.3.2.5   The pq eyewear brand story

Similar to Mykita, pq by Ron Arad also includes a variety of brand stories where the product plays on different levels (Fig. 6.21). The immediate impression from visiting the company website is a video playing on quirky, playful shapes, messages about being different meeting high tech. The website features a tagline: "See the world through our frame—See the world differently" [25]. This is complemented by an explanatory text: "Antiordinary is more than just a tagline: it's the mindset of pq by Ron Arad." Furthermore, the website introduces the three product lines, Ron Arad, technology, and a section where the customer can define his or her own sunglasses.

The company also makes a point in making it explicit that it is located in northern Italy.

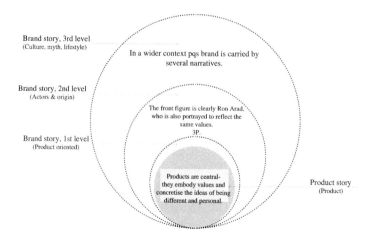

Figure 6.21   pq plays on all three levels.

Company headquarters are said to be in Susegana, near Conegliano, Italy, presented as a special place where Ron Arad's designs are brought to life: "From here π BY RON ARAD serves a global community of people who see beauty in difference nestled among hillside vineyards, in this inspiring location we have access to Eyewear's greatest minds and talents - from specialist manufacturers to leading technologists" [25].

On its home page, pq highlights a partnership with 3D systems, a company producing equipment for AM. AM is not in itself the main driver in the brand story. However, the company is built around the opportunities that AM offers.

Beyond the webpage, searching for pq online leads to blogs about 3D printing and various interviews with Ron Arad. Central messages concern thinking differently and that new technology is not an end in itself but may help create new opportunities. pq also attracted celebrities, and the company's products gained attention after Oprah Winfrey wore a pair of glasses on her show. Without doubt the most central part of pq by the Ron Arad brand story is Ron Arad himself.

6.4 Potential of Additive-Manufactured Products in Building a Brand

In this final section we will point at some of the key insights in these two cases, as well as look for potentials and challenges that arise when introducing AM in building a brand.

6.4.1 How They Used AM in Building a Brand

In the two cases investigated, the additive-manufactured products were used in building the brands in various ways. The Mykita brand emphasizes a creative environment through the Mykita Haus and the team behind, while the pq brand emphasizes the creative designer Ron Arad.

In the Mykita brand, AM was used as one of many technologies to allow them to meet their aim of making technical solutions that "must also be an aesthetic one." In the overall brand story, AM in itself was not important. However, it is important as a new technology that offers some benefits that they could use in making new eyewear. The manufacturing process was well documented on the web page with professional pictures (Fig. 6.22). In the Mykita Mylon eyewear they were interested in the material benefit AM gave (light, sturdy material) and this was linked with an area for use (sport/fashion). This leads to the fact that it is not important for them to make a statement in the products themselves that these are additive-manufactured. As new technology and collaboration are drivers in the development of new offerings, the next step for them has been to launch the My Very Own service. In this they are collaborating with various partners to make the most out of the new technology and can offer a far more personal and customized service where the customers take part in the design of their "very own" glasses.

Figure 6.22   Mykita Mylon production process. Photo: Mykita.

In the Ron Arad pq brand the business opportunities that lay within AM seem to be a core reason for setting up the company in the first place. An explanation of the AM technology being using has a prominent place on the web page and pictures that clearly show how the production method gives a particular texture (Fig. 6.23). It is used for the advantages the technology gives from a business perspective, as well as making use of the potential in the form. The most prominent is that of 3 Ps, a process that allows the customers to design their own eyewear according to predesign templates. In addition, they explore the variety stemming from AM by holding some signature elements constant, which supported recognition, while allowing diversity within the product portfolio. Diversity was also constrained beyond specific signature elements, for example, by keeping variety within a theme (names inspired by subway stations).

In a brand where stories are about creativity and exploration, like the pq brand, how AM can be used to explore and be creative could be a product story. It could also be the brand story if the company deliberately sees this as the position and communicate this in its activities.

Figure 6.23   pq by Ron Arad. Credit: pq by Ron Arad.

6.4.2   Opportunities in New Production Techniques

Moving away from mass to small-scale production is a benefit, which means that the company with less could produce limited editions and experiment with forms as it is less expensive to change. This may make it easier to obtain customer insights in the development process as the customer can experience a prototype of a product. With AM it is less costly to change standardized options, for example, according to seasons. It may also make it possible to have a higher level of customizations than in conventional manufacturing. A higher degree of customization opens up new ways of thinking about the product's role in building a brand. Brands that are product oriented and would like to offer an innovative experience have with this technology an opportunity to create strong customer experiences. The experience the customer has with the brand when codesigning the eyewear becomes the impression the customer has of the brand. This could be what the customer chooses to share with other fellows. AM opens up an opportunity for creating new ways for customers to interact with the brand in the manufacturing of the products.

Novel ways of manufacturing may in some cases itself shape the character of a product enough that even if a variety of versions are made, they will come off as a category in its own right. This is to some extent the case with Mykita's initial sheet metal eyewear, in which the material gives the product a very distinct expression. However, the character emerging from AM does not in itself differentiate pq and Mykita from other eyewear.

Another interesting new challenge is cobranding when two or more go together to create together. In both of the two brands explored in this chapter, partnership is key to making use of AM. The partners behind this were listed and presented and thereby are an important part of the story that these brands tell.

6.4.3   Form Freedom and Brand Development

With AM, many of the previous constraints found in conventional production change. The designer might have greater form freedom, which offers opportunity to create distinct forms. AM will give the design team an opportunity to design form based on the intended user experience and with a certain identity, rather than being limited by how it should be produced. In this way it is easier to make a distinct form, different from the competitors, and gives the customer a clear feeling of what the brand is about. If the design team manages to identify some element that can carry the brand, such as the spine-inspired hinge of pq, AM may then be central to how design elements and products communicate the brand.

A designer may engage with the novel opportunities so as to explore its possibilities, that is, pushing the opportunities AM offers may almost have an intrinsic value. The ranges of products at Mykita and pq may reflect this—using form elements that would have been difficult to realize in other manufacturing techniques. Whether AM becomes visible in the product depends at least in part on how designers and brand managers choose to relate to it. Seen as a means to realize products that are made to measure, the difference may be in fit rather than overall style. AM may from this perspective be a means to realize potential that was previously constrained.

Other aspects are time and skill set. Will the designers have time to design new forms, or will they have to build on previous forms to save time? When is it beneficial to start a new design, and when is it better to build on previous concepts? The designer team's knowledge, influence, and skills also will affect how AM is used. Are the team members curious about how this can help them to have a dialog with the users? Do they communicate the brand more efficiently, or are they more fascinated about the material quality and the opportunity to experiment with various shapes? Their design intentions, along with their knowledge and skills, will also inform how they use the new technology. Are they using new technology by replicating what they know from the past, or do they also have the skill set and knowledge to make use of new opportunities emerging with AM? To make decisions on this a higher awareness of how the product is part of building the brand is needed, in addition to what makes the product look the way it does today and what the benefits will be in a branded context to change it.

At the same time as AM opens up new opportunities, it also raises questions. Form limitations in branding could also be a benefit, promoting continuity. This will help them to communicate what they are about and in which product category they belong. When there is an opportunity to change every season, how do you still design-in the feeling of the brand in every product and create a coherent product portfolio? It will be important to define what parts of the form need to remain constant and what parts could be changed. The two frameworks offered in this chapter are introduced to help the design team to plan this.

In using AM to meet brand-building goals it will be important to understand the drivers in order to understand the design space and how to design an intended user experience and make this concrete (by design elements and interactions). This is the first step to better control further development of the product and how new technology such as AM is changing how products and brands are perceived. Decisions made (consciously and unconsciously) will at the end affect the final product.

6.4.4   Potential of Disruptive Stories

The cases show that there are various ways AM could be used in building a brand. In AM lies the potential for many brands that seek these as their stories. It could be the story about how the product is cocreated with the customer, or it could be about customization or how the design can change every season by use of AM. It can also be about the team behind the scenes that made this happen, such as the designers that had a particular interest in this technology. It could be used to produce products for a specific lifestyle or with a distinct identity. AM also currently comes with associations to advanced technology, something that may be an asset or burden relative to brand values.

One of the challenges for companies that would like to use this technology today is that the role of the products and the user experience offered by these products are poorly defined. The meeting points between branding, design, and manufacturing are less explored. There are also a variety of drivers behind companies. For these companies many of the aspects introduced, such as collaboration, organizational structure, and past performances, may be hindrances in exploiting AM, and they may be stuck in conventional manufacturing and management principles.

AM allows a higher dynamic in the design process and is a great tool to have in organic development of the brand. When the company has a good understanding of what the intended message is, which elements communicate the message, and how customers understand these messages, it can make good use of this as a tool to define how the product can be used to convey and build the brand. In AM it opens up for a far more intimate relationship with the customer, where it is not only the products themselves but the process of interactions that could be part of the brand experience.

A new technology such as AM, with the new opportunities that it gives, could change the category over time. In the beginning, it might be just the odd newcomer that is experimenting with it. However, once this newcomer has succeeded, bigger players will seek to use it; it can change the category rules and influence the design in every company. The stories about Mykita and pq offer lessons from both branding and AM points of view. They have by using this technology introduced not only new designs but also new user experiences and behaviors. This indicates that AM could be a driver of change that could open up different behaviors and lead to social or economic changes.

References

1. 2020Europe (2015). Ron Arad: My idea of design. Available online at http://www.2020europe.com/en_en/magazine/fashion_and_lifestyle-1/interview-6/ron_arad_my_idea_of_design-30.html (accessed on May 16, 2016).

2. Andersson, T., and Warell, A. (2015). Aesthetic flexibility in the management of visual branding. *Procedia Manufacturing*, **3**.

3. AMA (1995). American marketing association dictionary. Available online at https://www.ama.org/resources/Pages/Dictionary.aspx?dLetter=B (accessed on May 16, 2016).

4. Karjalainen, T.-M., and Warell, A. (2005). Do you recognise this tea flask? Transformation of brand-specific product identity through visual design cues. In *Proceedings of the International Design Congress, IASDR 2005*, Taiwan.

5. Killi, S. (2007). Custom design: more than custom to fit! *Virtual and Rapid Manufacturing: Advanced Research in Virtual and Rapid Prototyping: Proceedings of the 3rd International Conference on Advanced Research in Virtual and Rapid Prototyping*, Leiria, Portugal, 24–29 September, 2007 (s. 777–783). London: Taylor & Francis.

6. Cantarini, G. (2015). Ron Arad on fashion. *Hunger*. Available online at http://www.hungertv.com/feature/ron-arad-fashion/ (accessed on May 16, 2016).

7. Christensen, C. (1997). *The Innovator's Dilemma: When New Technologies Cause Great Firms to Fail.* Boston, MA: Harvard Business School Press.

8. Danforth, C. (2015). MYKITA's Moritz Krüger on celebrity endorsement, high fashion and more. *Highsnobiety*. Available online at http://www.highsnobiety.com/2015/05/05/mykita-mortiz-kruger-interview/ (accessed on May 18, 2016).

9. DeAcetis, J. (2013). The capitalist touch: the dedicated vision of MYKITA. *Forbes/Lifestyle*. Available online at http://www.forbes.com/sites/josephdeacetis/2013/04/01/the-capitalist-touch-the-dedicated-vison-of-mykita/#7ea3ce492b7b (accessed on May 18, 2016).

10. Etherington, R. (2012). A-frame by Ron Arad fo pq. *DeZeen Magazine*. Available online at http://www.dezeen.com/2012/06/15/a-frame-by-ron-arad-for-pq/ (accessed on May 16, 2016).

11. Dr Martens (2015). A history of rebellious self-expression. *Dr Martens Homepage*. Available online at http://www.drmartens.com/us/history (accessed on October 2, 2016).

12. Gillilan, L. (1999). Interiors: well-tempered rooms. *Independent*. Available online at http://www.drmartens.com/us/historyhttp://www.independent.co.uk/arts-entertainment/interiors-well-tempered-rooms-1095512.html (accessed on October 2, 2016).

13. Jordan, P. (2000). *Designing Pleasurable Products: An Introduction to the New Human Factors.* London: Taylor & Francis.

14. Holt, D. (2004). *How Brands Becomes Icons: The Principles of Cultural Branding.* Boston, MA: Harvard Business School Press.

15. Hestad, M. (2013). *Branding and Product Design: An Integrated Perspective.* London: Gower Applied Research.

16. Hopkinson, N., Hague, R., and Dickens, P. (2006). *Rapid Manufacturing: An Industrial Revolution for the Digital Age.* Chichester: John Wiley & Sons.

17. Killi, S. (2007). Custom design, more than custom to fit!. *Virtual Modeling and Rapid Manufacturing, Advanced Research in Virtual and Rapid Prototyping*, Vol. 3.

18. Madaboutspecs (2016). *Homepage.* Available online http://www.madaboutspecs.co.uk/mykita (accessed on October 16, 2016).

19. Mellery-Pratt, R. (2015). A closer look at the $13 billion premium eyewear market. *Business of Fashion.* Available online at http://www.businessoffashion.com/articles/intelligence/a-closer-look-at-the-13-billion-premium-eyewear-market (accessed on May 16, 2016).

20. Mykita Mylon (2016). The future look of sports. *Lookbook 2016.* Germany.

21. Mykita (2016a). Homepage. Available online at https://mykita.com/en/collections#layer:/en/collections/no1 (accessed on May 18, 2016).

22. Mykita (2016b). Homepage. Available online at http://mykitamylon.com (accessed on May 16, 2016).

23. MyVeryOwn (2016). Homepage. Available online at http://myveryown.com/en (accessed on October 16, 2016).

24. O'Connor, C. (2014). 5 Men's eyewear brands you should know. *Fashionbeans.* Available online at http://www.fashionbeans.com/2014/5-mens-eyewear-brands-you-should-know/

25. pq eyewear (2016). Homepage. Available online at http://www.pqeyewear.com/#about@ (accessed on May 16, 2016).

26. Tess (2015). Introducing D-frames, Ron Arad's $800 3D printed PQ eyewear sunglasses. *3ders.org.* Available online at http://www.3ders.org/articles/20151031-introducing-d-frames-ron-arads-3d-printed-pq-eyewear-sunglasses.html (accessed on May 16, 2016).

27. Tiger, L. (1992). *The Pursuit of Pleasure.* Harvard: Transaction.

Chapter 7

A Tale of an Axe, a Spade, and a Walnut: Investigating Additive Manufacturing and Design Futures

Andrew Morrison

Centre for Design Research, Oslo School of Architeture and Design, Oslo, Norway

andrew.morrison@aho.no

7.1   Prelude

Askeladden is a central figure in Norwegian folk tales and features prominently in the classic collections compiled by Asbjørnsen and Moe. In English Askeladden is known the Ash Lad (or at times Boots). In many settings, he is an astute observer, often remaining in the background as it were, keeping the embers warm by blowing on them until they burst into productive flames. Often, he is characterized as the reluctant, seemingly naïve, yet ingenious, youngest of three brothers. He waits patiently in the wings. He then acts adaptively, rather like a designer, appropriate to need and context, succeeding where others have failed, offering alternatives that have eluded others. In showing a way forward, in many folktales

Additive Manufacturing: Design, Methods, and Processes

Edited by Steinar Killi

Copyright © 2017 Pan Stanford Publishing Pte. Ltd.

ISBN 978-981-4774-16-1 (Hardcover), 978-1-315-19658-9 (eBook)

www.panstanford.com

his narratives operate allegorically but also materially. For design and design research we are easily able to read him as the designer; as the trickster; as an adaptive, resourceful, and generative thinker who worked with materials and contexts at hand to shape alternate, better futures.

In this chapter, Askeladden appears, disappears, and reappears textually, as he does in other stories in the folktales collection. I include him as a narrative and conceptual device for unpacking some of the future design-centered relations between product design (PD) and additive manufacturing (AM). In this regard, Askeladden contributes not merely as a provider of solutions but as a foil to assumptions and expectations of relations between PD and AM. This canny figure from Norwegian historical popular culture makes it possible to discursively investigate potential proximities between PD and AM. This includes the wider implications for transdisciplinary design practices and analysis concerning emerging technologies that are presented by way of market-driven promotion that centers on their presentation as popular cultural and accessible tools. The axe, spade, and walnut are artifacts that he finds, gathers, and deploys to meet the challenges of a future scenario partly identified in the present. The allegory of the folktale provides us with a humanities-oriented narrative and discursive view on unpacking relations between PD and AM.

In this context, the figure Askeladden is not interpreted literally but through an allusive and altered narrative that is offered as a means to unpack assumptions and expectations about AM and its uptake in the domain of PD. He is deliberately taken up as a ruse to accentuate the role of the trickster in design, a figure who is seemingly adept at pulling answers and possible futures out of his hat, or his rucksack, as we will see. This text thus draws on humanities inquiry in relation to design, products, and digital fabrication and especially narrative, with its attendant features of personas and scenarios that tend to be applied more functionally in pragmatic user-centered accounts of design practice. The chapter also attempts to tease out the unnatural, noncorrelational, and "fantastic" potential of narrative to provide us with a space and a performative mode of inquiry for shaping an alternative design discourse concerning making, technology, artifacts, and futures.

In the chapter I, therefore, also link humanities views with ones from the field of futures studies that is typically concerned with issues to do with more instrumental or deliberate approaches to strategic planning and strategy. Recently, interest has turned to investigations of foresight and anticipation in which narrative features. However, design practice and research seldom appear directly in futures studies, nor do they appear in the domain of foresight and connections to science, technology, innovation, and strategy (e.g., Ref. [57]).

There is a need for design, especially the force of technology in AM and its impact on practices and perceptions of PD, to inquire more deeply into what that the future might be [98] and how it is currently analyzed in futures studies. There is also the need to unpack relations between future views, and emerging technologies may be understood more critically by drawing on earlier engagement with the digital [60], as well as futures and AM [13].

Writings on AM rarely focus on design, mostly material, tools, and technologies. I see a strong need for the views of designers and designer-researchers to take up emerging technologies such as AM and to devise descriptive and analytical discourses of their own on the basis of their specialist expertise and experimental and creative collaborations. Here, too, there is room for social science critiques of technologies and also those of humanities. In this chapter I make a specific humanities turn by invoking the aesthetics of the baroque as a means for situating the additive, generative, and emergent in design and futures and their relations to connecting PD and AM.

The baroque offers us with a well-established set of characteristics or traits that we can usefully extend to the potential cross-connections between design futures and PD views on AM. The baroque was an epoch that was characterized by breaking given frames and working with "excess" in order to manifest materiality and representation, as will be elaborated. It allows us to take up the "additive" and construction of AM as digital fabrication in the context of PD, and this I do through narrative.

7.2 Queries

Across this book we have seen a variety of design-based research accounts that have taken up some of the key matters concerning relations between PD and AM. This chapter draws on these partner ones. It attempts to situate the diverse contributions within a wider perspective on design research that addresses matters of emerging technologies, exploratory practice, and transdisciplinary analysis. It does so in the context of design that is increasingly practiced and analyzed as a medley and hybrid of products, interactions, services, and systems [50, 83] and increased complexity, and the need to unpack relations between design, change, and conditions within an accelerated pace of technological development is occurring (e.g., Ref. [44]). Most of the research literature on AM, including the small body of work on PD, does not enact such a transdisciplinary mesh of relations and distinctions. Such mixed moves are hard to shape and share, and they are demanding to perform. This is precisely because they need to take up complexity with emergence in design inquiry; they also ask that we venture into developmental yet also generative relationships between domains that are often relegated to disciplinary silos.

PD itself is undergoing considerable changes as it shifts from earlier crafts-oriented preindustrial construction, through mass production in the industrial age connected to the growth of consumerism, into a postindustrial zone where the intersections between making, demand, use, and value are being argued as being dynamic, customizable, and more fluid [28]. This is often presented largely as a matter of the malleability of materials and digital tools so that relations between making and consuming are reconfigured, with the "prosumer" projected as the new driver of motivated, just-in-time user of self-directed demand [77].

As many earlier chapters in this collection have shown, this is not necessarily so simple as a switch. AM is promoted and projected as having transformatory potential for both means of artifact production and modes of commercial work [43]. Printing on demand is one of the characteristics AM with potential for customization, yet design expertise and precision are still needed in the context of actual new design work [48].

What then are the futures of AM from a PD view? What might AM be understood to be when PD perspectives are taken up, as Killi [47] argues? But what might this mean when PD is seen as interfacing with other domains of design, departing from earlier separations of design, in the context of the Internet of Things (IoT) and the embedding of services and interactions in physical artifacts? What, too, might be some of the implications for PD if we move beyond some of the bounded framings of product development contained within the approach of new product development (NPD)? How are we to look beyond the promotional marketing of technodeterminists who naturally have a vested interest in forwarding the developments in tools and materials?

7.2.1   On Discursive Design

For many readers of PD and of AM, humanities views on design may be somewhat unfamiliar and perhaps remote from the discourses and practices of engineering and product development. Yet, they offer us partners in sense making concerning the inclusion and uptake of new technologies. Much of this has been tackled through technology critiques through the situated studies of science in the domain of science, technology, and society (STS).

Narrative, dramaturgical, and performative aspects of the enactment and articulation of design-centered technology application are often bypassed, and functionalist and instrumentalist arguments and approaches predominate that are characterized by tool and technology promotion and the presentation of uncritical determinist views. Such humanities views as the ones mentioned allow us to deconstruct both given and formative discourses of technology. Yet that also allows us to get inside how design-based knowledge and its material manifestations may be shaped and critiqued.

I, therefore, draw on speculative design [24] and, acknowledging critiques of its gallery and fictional setting and imaginary status, use the conjectural to pose conceptual, alternate frames for interpretation of technology that is extensively presented in factual, actual terms elsewhere. There is need for design, theoretically and in practice, to more fully engage with the speculative in ways that are not simply indulgent and self-referential, but to think, to problematize, and to suggest.

Design research, futures studies, and STS make little connection to the extensive work in the humanities in productively and interpretatively untangling and yet also building knowledge about our sociomaterial and cocreative practices. This is knowledge that is not only seen through traditional humanistic hindsight but may also be understood through what elsewhere has been termed a "prospective hermeneutics" [68]. Following Butler [16], this chapter is thus excessive in its performative enactment of a transdisciplinary view on relations between the emergent in design and AM.

This chapter is, therefore, not a text that will lead you to a dramatized destination. It will offer only a version of a formulaic traditional folktale. Yet, like its oral tellers have always done, the text will offer embellishment and diversions. It will work paralogically (not deductively, inductively, or linearly) in its rhetoric. Narrative works to focalize, as Bal [9] argues. It may help us distinguish between the vision through which elements are mediated and the identity of the voice that is verbalizing that vision. And there then has to be a twist in the anticipated narrative conclusion to the tale and the research account.

Overall, this is enacted though an essayistic articulation that gathers the key thematic connected to an assemblage of related research. It builds a heterotopical and multiontological view on design and AM that reaches toward an anticipation view [73, 74] on futures perspectives. Poli [72] asserts, rather formally and deliberatively, that anticipation refers to systems that involve decision making in the present on the basis of forecasts about matters that may one day occur. He sees this view as allowing us to synthesize cognitive strategies from forecasts and scenarios into a system view.

The essay as enactment works additively in "manufacturing"—compiling and composing—an alternate discursive design research rhetoric of design futures [66]. As Sayers et al. [85] write:

While, as with any technology, physical computing and desktop fabrication can be exploited and deployed for oppressive purposes (e.g., surveillance, warfare, privilege, or monopolization), they also allow scholars to build alternatives, construct what-if scenarios, and create what, until recently, they may have only conjectured.

In design research one of my core interests is how we work speculatively. By this I mean how we develop what is conceptual and reflected and position the knowledge means and forms that may ensue as part of how to go about thinking about design, practice, and critique at a time when attention is being given to new materialism (e.g., Ref. [12]) and its sociomaterial design views that entail the participative and productive [58]. In AM circles, this has been part of a wide-scale debate about how and what can be made, most visibly perhaps relating to the distribution of software for printing a functioning handgun and related claims to freedom of access and ownership (e.g., Ref. [104]).

7.3 "Problems"

Askeladden and his brothers turn up outside a castle where the king has given the challenge that an oak tree be felled. Whoever fells the tree and fills the well with water is promised half the kingdom and the hand of the princess.

The trouble with this specimen of oak is that not only is it challengingly sturdy for the best of strongmen and their axes but when chips do fly off its trunk, they immediately propagate new trees. This results in the king brutally punishing these seeming offenders to his wishes. He cuts off their ears in public humiliation. The princess, historically never presented as having views of her own that steer the narrative, looks on.

In this contemporary recounting of the folktale, however, the narrative and characters adopt turns of their own making and choice, though these are entangled in the given structure.

Tools, expertise, expectations, scenarios beyond comprehension, and gaps in knowledge and response. Challenges. Wholly design- and design-research-centered features in need of creative responses. What's the parallel to AM's powerful demands and expectations? Will solutions simply be sought? Or will there be a less linear narrative progression and outcome? What are the possible design-rich spaces for exploring the coecologies of PD and AM?

7.3.1 Design, Narrative, Futures

In short, the twenty-first-century technodeterminist projections of AM may be understood as a continuation of a technodeterminist logic encountered across the twentieth century (e.g., Refs. [27, 37]) on two influential, conjoined axes. The first is a future positive drive powered by what has been constructed as a progression of technical development "married" to commercial interests.

The second, and more recent in an age of distributed and social media, is fueled by our fascination and engagement as consumers and, increasingly, producers of digitally mediated artifacts and expressions. Below I address these two futures concerning AM through critical reflection on technology aspirations and by way of exploring some possible, potential, and plausible—but also importantly projective—aspects of design-based futures around expectation and anticipation.

The retold and repositioned narrative of an axe, a spade, and a walnut provides us with scenarios that make it possible for us to discuss what is expected. Its construction also gives us a designerly space to express what might be anticipated differently as given contexts may be maintained and, when slightly altered, different outcomes might be arrived at. Methodologically, I reference the use of scenarios typically applied in user-driven design research as a mode of participative design making and reorient them in a futures view (e.g., Ref. [75]) toward contexts of conceptual reuse, reappropriation, and repositioning. It is attention to how this operates as an articulation of a possible way to think forward and into emergent conditions, not functionalist solutions to a defined problem that is central.

Writing on relations between actor network theory and participatory design, Stuedahl and Smørdal [96] argue that this is a matter of becoming that entails negotiation and translation. In the next section I take this up by focusing on the essay as a rhetorical device open for the articulation of such possibilities and enactments. I intend to show that the humanistic essay can offer analytical design inquiry potential for elaboration of the relations between PD and AM that currently are largely unexplored.

7.4 Essayistic

Askeladden and his two brothers move away from the castle. They have seen the many failed attempts to hew the oak tree and to fill the king's well with water. Failure results in punishment, so they decide to go on their way.

The band of brothers begins its journey up into the mountains and an unfolding narrative of happenstance, serendipity, and clarification.

Askeladden knows he has to gather his wits about him and draw on the materials, situations, and potential that they encounter. But he is the youngest brother, so he needs to be patient as his siblings take pole position and act first and second as they encounter familiar artifacts. His brothers stare at the world in front of them. Askeladden knows there is more to the object and the materials he encounters, so he must be "crafty" in his choices.

This chapter is deliberately presented in the format of an essay that has been chosen craftily, that is a means suited to toggling between narrative, description, critique, reflection, and interpretation. It is a historical form, developed to realize argument, the marshaling of claims and illustration, less empirical and more philosophical in its origins. The essay is rarely used in design research, yet it offers a space for description with interpretation [89]. It further allows us to frame a number of conditions, contexts, and developments so as to discuss and better unpack some of the assumptions and potentialities of design-based inquiry and qualitative research. The essay is also not prevalent in AM or PD, so this format is chosen to make space for a discursive design framing of the relations between PD and AM—and their futures.

STS scholars write situated critiques of technology uptake; humanists may employ essays, narrative sweeps, and speculative voices about what could exist, informed by fact and located in knowledge of other research. This essay, therefore, takes up the prospective and anticipatory not predictively but heuristically. Here this means that indirect allusion and possible potential constituents and configurations of the two are addressed not in a problem-solution frame but in a possibility space as yet not fulfilled but one that bears generative, alternate, and unexpected but also likely outcomes that need to be negotiated and teased out through processes of dynamic engagement.

Here, heuristic is interpreted more in its philosophical sense of shaping and projecting as a device for perceiving and thinking than that taken up variously in psychology as a more direct means to decision making. My "essaying" splices the entertaining and educative stance of a moral tale and its structuralist elements with more poststructuralist ones centered on figurative reflexive discursive enactment. This, too, is an instance of discursive design; in this chapter, narrative is the key mode of enactment.

7.4.1 Narrative

Not long into their climb into the mountains, Askeladden and his two brothers hear the noise of wood being chopped. Askeladden decides to investigate and to his surprise sees an axe all on its own chopping away at a fir tree. The axe announces it's been waiting for the young man. He smiles in amazement and seizes it, wondering whether it might be of use in the future. The blade looks sharp, but the handle needs replacing.

Askeladden quickly knocks the head of the handle on a nearby stone, retrieves its gleaming metal wedge form, and pushes the handle and head into his backpack. It feels comfortable there, next to his body. Joining his brothers on the ascent, he sees their puzzled faces and briefly explains he found an axe; one of them calls Askeladden a magpie.

Narrative needs to be seen as a discursive mode that affects all semiotic objects [9]. In this chapter, I draw theoretically on the narrative work of Bakhtin and his core concepts of polyvocality, modes of address, and the role of exhibitionism of pastiche in attention to the carnivalesque [6–8]. This dialogical theory allows for the uptake of concepts of addressivity, ventriloquism, and polyvocality. They are included in work I have been engaged in on design fiction (e.g., Refs. [65, 66]) that draws on an unnatural or nonmimetic narrative [2] and recent writings on narrative inquiry (e.g., Ref. [4]).

Concerning digitization and Hollywood cinema, Ndalianis [71] argues that contemporary film and entertainment media in a digital age exhibit a baroque poetics. Her insights echo well with the arguments and features of transdisciplinary design presented above. Ndalianis [71] writes that it is the lack of respect for the frame that is central to her work connecting baroque aesthetics and contemporary mediated culture.

There would seem to be many possible parallels here with the transdisciplinary design mix of products, interactions, services, and systems, not only media and narrative. Seventeenth-century baroque aesthetics operated as a mode of theater [26]. In part a tactic of the postreformation catholic church to provide experiential engagement, it may be characterized as breaking out of the frame in literature, art, and architecture. This was achieved, for example, in visual distortion and perspectival illusionism, juxtapositioning, and contrasting of image properties and through attention to mediation that reached beyond the given and into excess, allegory, and flux [15].

Latin American variations [26] moved from such major moves to minor ones that were more subversive, postcolonial in their subaltern voices, and pitched to ironize existing power relations. It is the heterogeneous and distributed nature of the unfolding of such relations in the baroque between multiple viewpoints and acts of knowledge production that Law [54] sees as a resource for further application in STS. Such traits arise primarily from humanities-centered inquiry and narratively they allow us to work beyond fixed frames of scenarios, situations, and expectations to create alternate spaces for wayfinding via design rather than problem-solving notions of addressing the future and design.

My focus here on narrative indicates that there is considerably more room than is often acknowledged for perspectives located in the humanities where narrative functions as a cultural resource for thinking, knowing, and communicating. Similar to the dynamics of designing with AM, this is an additive, malleable, combinatorial, and coeval cocreational view that may be looped back as a resource for understanding relations between PD and AM.

From the perspective of un/natural narrative theory this is about suspending disbelief, venturing beyond literalism, and making imaginative creative leaps [67] to articulate and engage in the projection of alternate views [66, XX]. This is to take up the role of narrative as plastic and always participatively situated in the eyes, hands, bodies, and minds of audiences in shifts from the folktale to an experimental poststructuralist baroque narrative.

Asbjørnsen and Moe (1852), the collectors of Norwegian folktales, saw themselves as gatherers or compilers of traditions and thus as retellers, not merely custodians, of fixed narratives. These narrative custodians drew together oral storytelling cultures and practices [20, 21]. They were concerned, however, to present shorter forms designed to be memorable with identifiable actors and scenarios and structures. Such communicative affordance allows for adaptation and embellishment around a core narrative repository of assembled tales. Their work has served to reinforce the notion that storytellers are valuable cultural intermediaries, and in the case of the folktales these may mediate the views of core dynamics connected to the comedic, animal, and supernatural (e.g., Ref. [40]).

It is within this focalization that Askeladden features as a lingerer of sorts who surveys situations as design potential spaces and waits to obtain artifacts dismissed by others, only to relate these artifacts one to another, seeing eventually purposes and means to overcome adversity. He takes up the trope of adventure, but where adversity and repetition may reoccur across tales in which he appears, there is typically a return to the real world where poetic justice is completed.

7.5 Promotion

As they proceed, on up into the higher reaches of the mountain, the three brothers come across a rocky overhang and there they hear the sound of digging. Ever curious, Askeladden decides to go and see who is at work this high up. His brothers now guffaw with laughter at what they see as their naïve young brother. "Never heard a woodpecker?" asks the second-eldest one.

After a difficult climb, Askeladden comes across a spade shoveling away at rocks. He stands in amazement. Once more the tool speaks! "It's about time, I was beginning to wonder if you'd understood the script," says the spade. Askeladden laughs and picks up the spade and turns it around in his hands, marveling at this find. Happy to have yet another artifact for his collection, he dusts off some stone. The spade feels light as he ties it onto his rucksack.

It is a long way from the softer ground of the lower reaches of the kingdom. When he joins his brothers, they look askance at the elegant piece of design sticking out of his pack. Askeladden watches their faces, smiling to himself. Shaking his head in disbelief, the eldest brother asks, "Do I have a donkey for a young brother?"

For many designers and design researchers AM takes three prominent forms that impact on their practice, professional memberships, and positions within the wider economy.

The first is physical and tangible in the form of increasingly cheap desktop 3D printers and the spread of other more high-end ones with greater material printing capacities.

A second aspect is that of the software that makes it possible to conceptualize, render, command, and print artifacts. This is clearly a change in the processes of making in that rapid versioning, scalable alterations, and processes of reflective revision are possible on-site and may toggle between software, screen, printer, and the materialized artifact.

Third, though, is the less tangible, more persuasively constituted marketing of 3D printing. This takes two main forms, the first relating to desktop printers and the propagation of maker space and "prosumer"-oriented arguments and practices [80]; the second is wider and refers to the role of mass-scale production changes that are projected as solutions to the economic fallout of the global financial crisis, especially in business circles in the US and yet also "folk-politics" as a response of the left to "postcapitalism" [93]. Let's briefly return to the closing decades of the past century.

1984, Apple, Orwell: Technological promise meets dystopian narrative meets future consumers. Apple launches its groundbreaking filmic advert for the personal computer, alluding to the fictional work of George Orwell projecting the rise of mechanically driven societies and repetitive, compliant human labor. As designers, educators, researchers, and consumers we have repeatedly experienced claims and persuasive offerings—and at times bold promises—about the transformative power of new and emergent technologies.

This is no less the case with the promotional discourses and strong claims made for AM [41]. The postindustrial society will be revolutionized through the application of the desktop 3D printing technology [32]. We enter a fourth Industrial Revolution [81] it is claimed. Mass production lines will be forever altered by the implementation of computational tools in "industries of the future" [82] and makers are to be part of this democratized desktop transformation of work [3] and thus design, through a new mode of "fabrication" [56].

Makers spaces, fabrication laboratories (fab labs), hacker zones, and 3D printing on demand are now features of a new mode of citizen production, collaboration, and distribution where anything, it is claimed, can be made [33] and where manifestos and projections of productive futures abound [39] and as the title of one recent book claims, *3D Printing Will Rock the World* [42]. However, we are not alone in our workspaces and hacker collectives, as the march of the robots is not to be heard far off in the distance; its foot soldiers are already dedicatedly at work as we sleep and fabricate [5]. Artificial intelligence (AI) has already ushered in an unstoppable machine intelligence that will render human participative labor ancillary to data-driven logics [30, 52].

Technology determinist claims are prevalent within innovation, business, and management circles. They are one of the responses to a reconfiguration of the complexity of contemporary work and the changing character of production, delivery, and distribution of goods and related services. We encountered similar assertions not so very long ago when in the mid-1990s the commercial Internet boomed and digitalization was fêted as forever transforming literacy, work, and leisure. Since then, many developments have indeed occurred and digital technologies are richly implicated in our daily and increasingly mobile-mediated communication, as work and leisure operate via networks and are influenced by them [11, 45]. Yet, as Morozov [63] has argued, technology is not the solution to cultural, economic, or social needs.

Yet, subsequent to the global financial crisis of 2008, the predictive, determinist claims for AM and AI have continued to grow precisely at a time when the very substrate and substance of monetism have themselves been under immense duress and reconfiguration. Indeed, some would say, this co-occurrence of technology-centric claims for economic revival needs to be seen as appearing at a time when the interests of the absurdly rich and the force of conglomerates have re-engineered their own accumulative interests at the further cost of others least able to provide routes out of poverty and exploitation. Technology is promoted as the route to helping highly developed economies to paint their way out of the corners of economic demise.

These claims need to be seen in relation to the underlying conditions they in effect erase: what goes unmentioned in these articulations are the deliberative and identifiable practices of offering bad loans, exercising corrupt and risky speculative financial practices, and exploiting pension funds as collateral, to mention a few of the features and practices of the so-called late capitalist economy that has not served, as countless experts and citizens now themselves assert, wider societal and democratic interests [79].

The claims for emerging technologies, now also transferred into "smart" cities and smarter homes, are, however, being met with critique [18, 59]. In part, this critique appears as the overarching claims are revealed to be weaker than their claims and their assertions are undermined through design and cultural practice. Birtchnell and Urry [13] offer a clear critique of claims for AM futures, referring to many design projects and even speculative approaches; however, design does not feature strongly in their surveys or analyses. We need, as Ratto [77] claims, a mode of critical making in which design features more prominently.

7.5.1 Intersections

Now much higher up into the mountains, Askeladden and his brothers pause for a rest beside a stream. Askeladden looks around him and wonders where all the water is coming from. In unison, the brothers ask him if he's never heard of a spring. Askeladden, not one to take things for granted, sets out to cure his curiosity. He follows the stream uphill until it grows smaller and smaller. Eventually he sees that it originates in a giant walnut of all things. Askeladden is amazed at this sight, pleased he made the effort to climb so far.

A walnut! And it speaks to him, telling how it has been trickling the water down the mountain side and waiting for his arrival. Askeladden stands quite still, his unknown ascendant quest now nearly fulfilled. He wishes to take this remarkable artifact with him, natural but fantastic, generative yet fueled by some magical source (code). He looks around him for some material to support his growing idea of what to do in the future and sees a verdant bank of moss beside the spring, He pulls away at it and plugs the hole with moss to stop the flow of water.

Askeladden returns to his brothers. The elder two now laugh out loud at the growing absurdity of their young brother's finds. Askeladden laughs gently, feeling a wave of warmth flow over his whole body, instantly energized. The brothers finish their packs of sandwiches. Askeladden begins to sense there is much more ahead of him. He cannot quite see what it is, but a fuzzy narrative shape is beginning to form in his mind as to how future scenarios might play out.

While recent research publications on AM have begun to shift away from more ostensive technologically and commercially centered approaches to AM and offer more critical, situated views from the human sciences (e.g., Ref. [28]), design is seldom mentioned. Missing are perspectives from ways designers and designer-researchers engage with emerging technologies and how they situate AM in relation to knowledge and expertise derived through design practice and analysis. Design-located knowledge and reflection in and on action in design-centered AM—on emerging technologies, changing materials, relations between the physical and digital in terms of tools and creative processes, and production methods—shift the focus toward ways AM may be understood creatively, innovatively, and generatively.

In many respects, this is a design-futures-making view [105] in that it connects aspirations and expectations with materializations through experimental inquiry. Increasingly, the boundaries between design domains are blurred as materials and products, interactions, and services are interconnected in the making of artifacts, the provision of processes, and the articulation of interactions. Research on AM has yet to address these intersections and needs to shift its attention away from claims for reconfiguring and enhancing postindustrial production and work practices and delve more deeply into ways design-centered inquiry may provide actual yet exploratory realizations of the application of AM technologies and tools (see Chapter 4) beyond earlier arguments and research on materials and infrastructures [41] and social change (e.g., Ref. [77]).

In their book *Divining Digital Futures* Dourish and Bell [23] present a critique of technodeterminist views of the digitally shaped and distributed character of our networked and socially mediated lives. AM does not feature in this critique nor in their suggestion that we pay attention to the emergence and situated uptake of the digital in a design and socially communicative frame. As Kempton mentions in Chapter 2, AM needs to be seen in relation to prior critiques of technology.

7.6   Foresight

The brothers return to the castle grounds. They recall the king's challenge that half the kingdom and the hand of the princess are the reward for the one who can fell the giant oak and fill the well. The tree has in the meantime doubled in size. And the same problem remains: for every swing of the axe, two chips fly from the tree.

Askeladden surveys the castle, the king, and the assembled crowd—an intimidating performative scenario. It's the moment for a convergence of thoughts and practical knowledge. He has gathered his material affordances and his epistemic artifacts and now intends to exercise his expertise over and through them. There is no room for error. The king and his subjects are filled with expectation: might someone finally complete the brief? No one has been able to predict their success.

The elder two brothers are fearless. They step forward and fly at the tree trunk with their axes but to no avail. They face the same fate as others who have failed: the king banishes them to a remote island and cuts off both their ears. The monstrous tree intimidates one and all.

Askeladden feels a shiver run down his spine. He has seen numerous heads roll as it were, their ears shorn off to the skulls by the king's minions. He stands stone-still and breathes in deeply, concentrating on his earlier intention to see his way through the problem and the problems others have created by their practices and repeated failures.

Askeladden recalls his thoughtful forward-reaching pause on the way up into the mountains—not a vision but an open willingness to develop foresight. Right now it's his embodied actions, not his conceptualization, that will reveal whether he's managed to foil the traps in the task. He gathers up his concentration and steps forward, takes off his rucksack, and feels the force of what is about to prevail take hold of him.

So if there are these technodeterminist economic predictions of the promise of AM, how might we look more skeptically and prospectively at the same time at futures of a PD view on AM? One possible direction is to connect to otherwise largely separate domains of knowledge building: design research and futures studies. Both tackle matters of what lies beyond the immediate here and now in their reach toward alternate, better worlds. However, for design, this is to leave some of its earlier, more positivist framings in the strive to ensure design be taken seriously as a science. For future studies this is to sidestep its often policy and strategic planning orientations and to move into less predictive framings.

The development of foresight has been one main feature of strategic approaches in futures studies. Slaughter [91], one of the leading figures on foresight, has sought to offer nondystopian views of the future that involve social engagement in foresight processes that are dialectically realized through the interplay of an expanded futures frame and experience (see also Ref. [61]).

Voros [101] took up more dynamic-oriented aspects of foresight in proposing a generic process view that allows foresight to be tailored to context, needs, and commissions in organizations with pre-existing strategy development and planning and via praxis to close the loop in action research. It is now over a decade since Voros [102] proposed what he termed a generalized layered methodology and an annotation of the futures cone (earlier developed by Hancock and Bezold). This cone usefully provides us with a schematized view on the relations between the possible, plausible, probable, and preferable in design. However, in my view what is missing concerning design futures is the projective and a multiontological view.

7.6.1 Scenarios and Futures

Arguments for connecting design research with futures studies have begun to gain some traction [88], though in many respects links are still tenuous and the role of design is still largely to be professionally and academically framed [19]. Selin [87] outlines ways collaborative initiatives may be developed and reported as a means to opening out constructive, collaborative spaces for connecting art, design, and strategic foresight. The application of scenarios has been central to strategic foresight work. It has been prominent in attention to the methodological that is also gaining interest (e.g., Ref. [35] and extending to organizations [99] and to participatory processes in qualitative inquiry [46].

The predominant view has been that scenarios are applied as devices to project today's views and expectation into likely and predictable contexts in the near future. Naturally, such uses have their place but they are bound by practices and perspectives in planning and business strategy and not driven by modes of inquiry that are more speculative and draw more on the contingent, emergent, and putative and open out possible, prospective design spaces.

What remains is for fuller distinctions and linkages to be made between the policy, strategy, business, and management views and analyses of foresight and one's generated form within the design domains that entail creativity, invention, and innovation [22, 34]. So, too, is there need to unpack what is meant by foresight when it moves through what are being termed "networks of innovation" [100] toward cultural, communicative, and mediational views, not merely planning and policy [66].

PD and AM clearly have a space to elaborate together in building foresight views that connect, for example, the IoT with contexts of possible and potential use, and also speculative and near-future imagined ones. These are nonliteral, marked by their playful forward-reaching character and motivated to foster thinking about the future and not merely predictive arenas for determined action.

In this regard PD's relationships to AM may be taken up as one way to work between the established in design, drawing on its longest professional formal academic legacy from industrial design to NPD, and to connect it to the forward-looking aspect of futures studies where accent is on the anticipatory, emergent, and potential. In making a linkage to futures studies, PD may thus also bring an alternate framing of futures to the discourses of AM. There is great potential in referring to research into scenarios in futures work, such as that reported across three very different domains and locations by Ramirez et al. [75] as a means to develop novel, challenging, and prospective methodological advances for transdisciplinary inquiry.

7.6.2 The Fictive and Nondeterminist Futures

Futures studies, however, do not look very deeply into the sociomaterial and sociotechnical discourses of emergent tool, technologies, and platforms. At the same time it could also be said that such views that are advanced in domains such as STS have not engaged with the body of research in futures studies. Perhaps this may be attributed to its focus on policy, strategic planning, and scenarios and origins within systems thinking and the influence of cybernetics and less on the social constructions of STS views.

Taken together, futures studies and design may present us with potentialities that may be centered around design-focused terms already in use, such as aspiration, anticipation, and projection [19].

How are we to get our heads around nondeterminist design futures when emerging technologies are always likely to be promoted as part of their assertion, novelty claims, and market-driven persuasive force?

One way of looking into relations between design and futures inquiry is to look into the fictive. This book has many chapters on the factive; this chapter offers ways that foresight may be taken up narratively to string together these three thematics in a prospective hermeneutics of additive, situated, systemic communication design ecologies. Though there are arguments advanced in the futures literature for attention to be directed to narrative (e.g., Ref. [78]) and to plausibility in scenario process planning, they remain largely in terms of reviews of literatures and approaches and do not apply narrative tools, methods, and strategies rhetorically nor do they refer to the performative or design as a domain.

Recently Eidinow and Ramirez [25] have argued for attention to the aesthetic in storytelling and story making as a means of engaging and persuading participation in strategic planning and decision making; narrative is not taken up as a semiotic and cultural-oriented resource for engaging with emerging technologies in exploratory and critical ways in shaping a discursive design approach.

7.7 Reflections

The crowd beside the giant oak has swelled, and Askeladden knows that there is no one ahead of him. His time has come. He has to take up the challenge and to use his ingenuity to not only solve the matter at hand but offer an alternate future to the conundrum. Everyone else has failed to fell the giant oak. The king's challenge still stands: "Cut down the tree and fill my pond with water and you win the princess and half the kingdom."

Askeladden takes in a deep breath and steadies his nerves. He puts down his rucksack and opens it out. There he sees again the artifacts he has found and chosen on his journey. He needs to think combinatorially, like a designer. What tools are suited to what tasks, and how might they have an added value? To what end might their conjunction be put, and how will this prepare him for not just possible but likely futures?

What, too, lies ahead of him as he reconfigures his expectations and the plausible staging of his dramaturgical acts that are now poised like the axe he finds in his hands? He takes in a huge breath and then tears at the tree trunk. Chips fly but before they can touch the earth he's dug out a huge hole so that they cannot multiply on the surface but are buried deep below the surface.

Then as axe and spade continue their automated actions without him, Askeladden rips the moss out of the walnut and allows its stream to flow into the hole, drowning the trees generative powers, filling the well, and securing him half the kingdom and the princess. But this modern designer Askeladden looks up and instead of claiming his gendered prize, he simply shouts out that a new brief is to be announced by the princess on her own terms and in her own time.

Always keen to work conceptually, and to find a fit between tools, purposes, cocreation, and participation, Askeladden simply laughs, leaving the challenge in the air as the princess waves to him nonchalantly from the castle turret. She is free of her male key keepers.

The princess leaves the parapet and returns to her sea-green velvet daybed, from where she has been watching the scenario play itself out in the two-way mirror she printed earlier in the week. "Just a little design to help me to look forward and backward at the same time," she'd whispered to the digital assembler. The Anticpatrix is what she's called her "many looking mirror."

Askeladden had already left the gathering, keen to return to his studio and the slip of programming he snuck in while earlier watching the flow of water from the walnut. He'd had an idea about a future use for an additively manufactured product and was now on his way back to his developer team to crack its conceptual nut.

Not all the metaphors had worked, an unnatural linear narrative had partly provided him with foresight, and the performative enactment had been partly serviceable. He shrugged off his designer's mantle and decided to take the day off from design storytelling.

From a design perspective, one of the main challenges facing AM is that of increased complexity in relations between the technical, sociopolitical, and economic. This complexity is compounded by technology-driven views and promotional practices geared toward projecting brighter commercial and technologically framed futures [82] and an uncritical solution-driven worldview concerning the changing nature of work, innovation, and marketing in a postindustrial age. Added to this context is that design itself is a mesh of domains and disciplines that may themselves not be fully connected and demarcated to best appoint the activities of designing transversally and researching through critical practice.

7.7.1 Toward the Additive in Discursive Design

This chapter has examined what it is that designers themselves see as the key issues to consider, and to consider together, in addressing the potential and pitfalls of AM. While their views are informative at many levels, they exhibit a tendency of design-based work in emerging technologies to steer away from perspectives and practices from art-based inquiry that allow us to more fully attend to the aesthetic, the communicative, and the expressive. This is not to suggest that design-based AM explorations should be geared toward creative articulations as art but that techniques and modes of working with the materialities of AM may be understood as more than material experiments or the marking out of infrastructural potential.

This chapter is one tentative "expedition" into a possible route of opening out and connecting thinking about such potentials and spaces for thinking about design-rich narrative views on futures via a hybrid rhetorical mode of discursive design. There is a vast expanse of potential for its articulation via practice, through detailed design cases and contexts, and for the growth of situated critique and reflection between making and analyzing so that our futures—especially ones that link tools, techniques, technologies, and telling as relations between PD and AM—may be ones that are not programmed and planned but investigated through design "ecologically," systemically, and communicatively. PD and AM may turn out to be one under-rehearsed arena for realizing these linkages and distinctions.

7.7.2 Design Baroque Futures

What is needed is a multiontological space [1] that allows us to think across and between and within our given and emerging practices and our modes of making knowledge that is able to tackle and account for complexity and conjecture. Ndalianis [71] writes that "like the precious baroque mirror, culture and its cultural products nurture and reflect back on one another in a series of endless folds, producing reflections that fracture into multiple, infinitesimal pieces, which finally also comprise a single entity."

In the baroque, this was taken up in poetry as a means to escape the Ovidian myth of Narcissus, whose self-absorption with his own reflected image led to his death [62]. Might there be more in rethinking the baroque into the dynamic, complex times of today, into different design-gendered identity and performative and participatory spaces, where tools and technologies are playfully interrogated to develop new knowledge and contextual use values? Discursive design is one domain in which this may be enacted experimentally where the status of knowing is emergent, yet performative.

There are many manuals on AM and numerous claims from innovation and business writers that the future lies in digital fabrication. Yet our experience of earlier technology drives revealed that the digital interfaces with the actual. Further, our mediated and embodied selves move between the digital and the virtual. Excellent products still need to be designed just as new materials and processes of production need to be teased out and applied. What is clear is that the sculptural, tweakable, and 3D character of digital fabrication offers PD new sites of engagement in which it is the design of products that is primary but the technology does not perform conceptually or critically.

It is the designer's role—in realizing the conceptual and embodied form, connected to contexts of use and uptake—that now places PD squarely at the center of design-informed AM. This is not possible without research from within design and without technology critiques of the discourses of AM and without clearer notions of what sorts of futures research may achieve what sorts of futures if design is not included. Lack of attention to these matters may lead us to seem as jaded and complacent as Askeladden's brothers. As a consequence, we may otherwise miss many of the discursive and physical affordances that we may benefit from when gathering and interrogating in contexts of need and potential use.

7.8   Generative Visions

The princess holds up her "many mirror." She is able to look in multiple directions, yet the handle of her device is firmly within her own grasp. She spins the artifact, gleefully, finally free of male gazes as the dominant lens. She observes the mirror's many images reflected on others' faces, across the small lake on the roof of the castle where she has been developing a garden to feed the inhabitants. She is delighted at her many mirror's generative visions. She watches as the light glints on the leaded windows and images are refracted and diffracted back into the world through a diverse mix of lenses and surfaces.

The princess pauses. She knows the future does not actually exist tangibly, but she feels alive creatively. Now she is able to see what the future might be as the mosaic of meaning she performs changes in its processes, offering panes and planes of potential.

The princess knows she still lives in the baroque, but she sees it as a time of multiples, or possible spaces for thinking and enacting her thoughts. Ah, she sighs, it's the future baroque.

The princess places the mirror in its stand and returns to her reading of the poetry of the Spaniard Gabriel Bocángel y Unzueta. There, to her surprise, she reads a poem on how the myth of Narcissus (so stressed by her tutor as a narrative of how image and reality may be understood to be fractured) is broken through the mirror itself being fractured. This repositioning through the creative work of her own times spurs on the princess's thinking to one additional turn. What if the mirror might work to produce multiples not separates, so that it might show both time and space in different ways?

So it was that the princess decided to commission the best designers in the land to develop her conceptual device further. She invited them to add to its forms and functions and to see them as bound together in a coarticulation of means and medium. Its users would be able to direct the potentialities of the device and understand its means to open out to multiple meanings yet include specific forms directed to particular uses. Not all of these affordances would be given, but some of them could at least be anticipated and others would remain milky and shadowy visions just beyond focus.

Acknowledgments

My thanks for inputs and insights to Dagny Stuedahl, Angeliki Dimaki Adolfsen, Henriette Killi Westhrin, and Sandra Kemp. This would not have happened without the invitations, ingenuities, and critical engagement of Steinar Killi and William Kempton.

References

1. Aaltonen, M. (2009). Multi-ontology, sense-making and the emergence of the future. *Futures*, **41**, 279–283.

2. Alber, J., Skov Neilsen, H., and Richardson, B. (2013). *A Poetics of Unnatural Narrative*. Columbus: The Ohio State University Press.

3. Anderson, C. (2012). *Makers: The New Industrial Revolution*. New York: Crown Business (Kindle edition).

4. Andrews, M., Squire, C., and Tamboukou, M. (2013). *Doing Narrative Research*. London: Sage (Kindle edition).

5. Armstrong, S. (2014). *Smarter Than Us. The Rise of Machine Intelligence*. Berkeley: Machine Intelligence Research Institute (Kindle edition).

6. Bakhtin, M. (1981). *The Dialogic Imagination: Four Essays*. Holquist, M. (ed.), Emerson, C., and M. Holquist (trans.). Austin: University of Texas Press.

7. Bakhtin, M. (1984). *Problems of Dostoevsky's Poetics*. Emerson, C. (ed. & trans.). Minneapolis: University of Minnestota Press.

8. Bakhtin, M. (1986). *Speech Genres and Other Late Essays*. Emerson, C., and Holquist, M. (eds.), McGee, V. (trans.). Austin: University of Texas Press.

9. Bal, M. (2001). *Looking In: The Art of Viewing*. Amsterdam: OPA.

10. Balsamo, A. (2011). *Designing Culture: The Technological Imagination at Work*. Durham: Duke University Press.

11. Benkler, Y. (2006). *The Wealth of Networks: How Social Production Transforms Markets and Freedom*. New Haven: Yale University Press.

12. Bennett, J. (2010). *Vibrant Matter: A Political Ecology of Things*. Durham: Duke University Press.

13. Birtchnell, T., and Urry, J. (2016). *A New Industrial Future? 3D Printing and the Reconfiguring of Production, Distribution, and Consumption*. London: Routledge (Kindle edition).

14. Bleecker, J. (2010). Design fiction: from props to prototypes. *6th SDN Conference: Negotiating Futures – Design Fiction*. Basle, October 28–30.

15. Buci-Glucksmann, C. (2013). *The Madness of Vision: On Baroque Aesthetics*. Athens: Ohio University Press.

16. Butler, J. (1997). *Excitable Speech: A Politics of the Performative*. London: Routledge.

17. Calloway, S. (1994). *Baroque Baroque: The Culture of Excess*. London: Phaidon Press.

18. Campbell, T. (2012). *Beyond Smart Cities: How Cities Network, Learn and Innovate*. London: Routledge.

19. Celi, M., and Morrison, A. (2017). Anticipation and design. In Poli, R. (ed.), *The Handbook of Anticipation*. Vienna: Springer (under review).

20. Christiansen, R. (1931). *Norske Eventyr (Norwegian Folktales)*. Oslo: Nordisk Kultur IX.

21. Christiansen, R., and Liestøl, K. (1931). *Norske Folksegner (Norwegian Folklegends)*. Oslo: Nordisk Kultur IX.

22. De Laat, B. (2000). Scripts for the future: using innovation studies to design foresight tools. In Brown, N., and Rappert, B. (eds.), *Contested Futures: A Sociology of Prospective Techno-Science*. London: Routledge, pp. 175–208.

23. Dourish, P., and Bell, G. (2011). *Divining a Digital Future*. Cambridge, MA: The MIT Press.

24. Dunne, A., and Raby, F. (2013). *Speculative Design*. Cambridge, MA: The MIT Press.

25. Eidinow, E., and Ramirez, R. (2016). The aesthetics of story-telling as a technology of the plausible. *Futures*, **84**, 43–49.

26. Eggington, W. (2010). *The Theater of Truth: The Ideology of (Neo) Baroque Aesthetics*. Stanford: Stanford University Press.

27. Feenberg, A. (1999). *Questioning Technology*. London: Routledge.

28. Ferdinand, J.-P., Petschow, U., and Dickel, S. (2016). *The Decentralized and Networked Future of Value Creation: 3D Printing and its Implications for Society, Industry, and Sustainable Development*. Switzerland: Springer International Publishing.

29. Floyd, J. (2012). Action research and integral futures studies: a path to embodied foresight. *Futures*, **44**(10), 870–882.

30. Ford, M. (2015). *The Rise of the Robots: Technology and the Threat of Mass Unemployment*. UK: Oneworld.

31. Gaver, B., Dunne, T., and Pacenti, E. (1999). Design: cultural probes. *Interactions*, **6**(1), 21–29.

32. Gershenfeld, N. (2005). *Fab: The Coming Revolution on Your Desktop: From Personal Computers to Personal Fabrication*. New York: Basic Books.

33. Gershenfeld, N. (2012). How to make almost anything. In Rose, G. (ed.), *The Fourth Industrial Revolution*. A DAVOS Reader: Foreign Affairs (Kindle edition).

34. Gheorghiu, R., Andreescu, L., and Curaj, A. (2016). A foresight toolkit for smart specialization and entrepreneurial discovery. *Futures*, **80**, 33–44.

35. Giaoutzi, M., and Sapio, B. (2013). *Recent Developments in Foresight Methodologies: 1*. New York: Springer.

36. Hand, M. (2008). *Making Digital Cultures: Access, Interaction, and Authenticity*. London: Ashgate.

37. Harraway, D. (2008). *When Species Meet*. Minneapolis: University of Minnesota Press.

38. Harraway, D. (2016). *Staying with the Trouble: Making Kin in the Chthulucene*. Durham: Duke University Press.

39. Hatch, M. (2012). *The Maker Movement Manifesto: Rules for Innovation in the New World of Crafters, Hackers and Tinkerers*. New York: McGraw-Hill (Kindle edition).

40. Hodne, Ø. (1984). *The Types of the Norwegian Folktale*. Bergen: Universitetsforlaget.

41. Hopkinson, N., Hague, R., and Dickens, P. (2006). *Rapid Manufacturing: An Industrial Revolution for the Digital Age*. Chichester: John Wiley & Sons.

42. Hornick, J. (2015). *3D Printing Will Rock the World*. North Charleston: CreateSpace Independent Publishing Platform.

43. Hoskins, S. (2013). *3D Printing for Artists, Designers and Makers*. London: Bloomsbury.

44. Ito, J., and Howe, J. (2016). *Whiplash: How to Survive Our Faster Future*. New York: Grand Central Publishing.

45. Johnson, S. (2012). *Future Perfect: The Case for Progress in a Networked Age*. New York: Riverhead Books.

46. Kaivo-oja, J. (2017). Towards better participatory processes in technology foresight: How to link participatory foresight research to the methodological machinery of qualitative research and phenomenology? *Futures* (in press).

47. Killi, S. (2013). Designing for additive manufacturing: perspectives from product design (unpublished doctoral thesis). Oslo School of Architecture and Design (AHO), Oslo.

48. Killi, S., Kempton, W., and Morrison, A. (2015). Design issues and orientations in additive manufacturing. *International Journal of Rapid Manufacturing*, **5**(3–4), 289–307.

49. Killi, S., and Morrison, A. (2016). Fea and 3D printing, the perfect match? *International Journal of Mechanical Systems Engineering*, **2**, 111.

50. Knutsen, J., Martinussen, E., Arnall, T., and Morrison, A. (2011). Investigating an "Internet of hybrid products": assembling products, interactions, services, and networks through design. *Computers and Composition*, **28**(3), 95–104.

51. Knutz, E., Markussen, T., and Christensen, P. (2013). The role of fiction in experiments within design, art and architecture. *Proceedings of NORDES 2013*, 9–12 June, Copenhagen/Malmö, pp. 341–348.

52. Kurzweil, R. (2015). *The Singularity is Near: When Humans Transcend Biology*. London: Duckworth Overlook.

53. Law, J. (2004). *After Method: Mess in Social Science Research*. London: Routledge.

54. Law, J., and Ruppert, E. (2016). *Modes of Knowing: Resources from the Baroque*. Manchester: Mattering Press (Kindle edition).

55. Lemke, J. (1998). Metamedia literacy: transforming meanings and media. In Reinking, D., McKenna, M., Labbo, L., and Kieffer, R. (eds.), *Handbook of Literacy and Technology*. Mahwah: Lawrence Erlbaum, pp. 283–301.

56. Lipson, H., and Kurman, M. (2013). *Fabricated: The New World of 3D Printing*. Indianapolis: John Wiley & Sons.

57. Loveridge, D. (2008). *Foresight: The Art and Science of Anticipating the Future*. London: Routledge.

58. Marres, N. (2012). *Material Participation: Technology, the Environment and Everyday Publics*. Basingstoke: Palgrave Macmillan.

59. Marvin, S., Luque-Ayala, A., and McFarlane, C. (2016). *Smart Urbanism*. London: Routledge.

60. McCullough, M. (1996). *Abstracting Craft: The Practiced Digital Hand*. Cambridge, MA: The MIT Press.

61. Mendonça, S. (2017). On the discontinuity of the future by other means: reviewing the foresight world of Richard Slaughter. *Futures* (in press).

62. Mercardo, L. (2016). Breaking the myth: Bocángel's new Narcissus. *Calíope: Journal of the Society for Renaissance and Baroque Hispanic Poetry*, **21**(2), 19–36.

63. Morozov, E. (2013). *To Save Everything, Click Here: Technology, Solutionism, and the Urge to Fix Problems That Don't Exist*. London: Penguin.

64. Morrison, A. (2010). *Inside Multimodal Composition*. Cresskill: Hampton Press.

65. Morrison, A. (2014). Design prospects: investigating design fiction via a rogue urban drone. In *Proceedings of DRS 2014 Conference*, 16–19 June, Umeå, Sweden.

66. Morrison, A. (2016). The seeds of design fiction. Conference keynote, *A Year Without a Winter*. Hosted by ISPRA and Arizona State University, ISPRA: Italy.

67. Morrison, A. (2017). Future north, nurture forth: design fiction, anticipation and Arctic futures. In Kampevold Larsen, J., and Hemmersam, P. (eds.), *Future North, The Changing Arctic Landscapes*, Ashgate: London (in press).

68. Morrison, A., Aspen, J., and Westvang, E. (2013). Making the mobile and networked city visible by design. In *Proceedings of Crafting the Future, 10th European Academy of Design Conference*, 17–19 April, Gothenburg.

69. Morrison, A., Arnall, T., Knutsen, J., Martinussen, E., and Kjetil, N. (2011). Towards discursive design. In *Proceedings of IASDR 2011, 4th World Conference on Design Research*, (CD-ROM), 31 October–4 November, Delft, Netherlands.

70. Morrison, A., Tronstad, R., and Martinussen, E. (2013). Design notes on a lonely drone. *Digital Creativity*, **24**(1), 46–59.

71. Ndalianis, A. (2004). *Neo-Baroque Aesthetics and Contemporary Entertainment*. Cambridge, MA: The MIT Press.

72. Poli, R. (2010). An introduction to the ontology of anticipation. *Futures*, **42**, 769–776.

73. Poli, R. (2014a). Anticipation: what about turning the human and social sciences upside down? *Futures*, **64**, 15–18.

74. Poli, R. (2014b). Anticipation: a new thread for the human and social sciences? *CADMUS*, **2**(13), 23–36.

75. Ramirez, R., Mukherjeeb, M., Vezzolic, S., and Kramerd, A. (2015). Scenarios as a scholarly methodology to produce "interesting research". *Futures*, **71**, 70–87.

76. Ratto, M. (2011). Critical making. In van Abel, B., Evers, L., Klaassen, R., and Troxler, P. (eds.), *Open Design Now: Why Design Cannot Remain Exclusive*. Amsterdam: Bis Publishers. Available at http://opendesignnow.org/index.html%3Fp=434.html (online).

77. Ratto, M., and Ree, R. (2012). Materializing information: 3D printing and social change. *First Monday*, **17**(7). Available at http://firstmonday.org/ojs/index.php/fm/article/view/3968/3273

78. Raven, P., and Elahi, S. (2015). The new narrative: applying narratology to the shaping of futures outputs. *Futures*, **74**, 49–61.

79. Reich, R. (2015). *Saving Capitalism for the Many Not the Few*. New York: Knopf.

80. Ritzer, G., and Jurgenson, N. (2010). Production, consumption, prosumption: The nature of capitalism in the age of the digital "prosumer". *Journal of Consumer Culture*, **10**(1), 13–36.

81. Rose, G. (2012). *The Fourth Industrial Revolution*. A DAVOS Reader: Foreign Affairs (Kindle edition).

82. Ross, A. (2016). *The Industries of the Future*. New York: Simon & Schuster.

83. Rossi, C. (2013). Bricolage, hybridity, circularity: crafting production strategies in critical and conceptual design. *Design and Culture*, **5**(1), 69–87.

84. Sayers, J. (2015). Prototyping the past. *Visible Language*, **49**(3), 156–177.

85. Sayers, J., Elliott, D., Kraus, K., Nowviskie, B., and Turkel, W. (2016). Between bits and atoms: physical computing and desktop fabrication in the humanities. In Schreibman, S., Siemens, R., and Unsworth, J. (eds.), *A New Companion to Digital Humanities*, 2nd ed. n.p.: Wiley-Blackwell, pp. 3–21.

86. Schatzmann, J., Schäfer, R., and Eichelbaum, F. (2013). Foresight 2.0: definition, overview & evaluation. *European Journal of Futures Research*, **1**, 15.

87. Selin, C. (2015). Merging art and design in foresight: making sense of emerge. *Futures*, **70**, 24–35.

88. Selin, C., Kimbell, L., Ramirez, R., and Bhatti, Y. (2015). Scenarios and design: scoping the dialogue space. *Futures*, **74**, 4–17.

89. Shields, D. (2010). *Reality Hunger: A Manifesto*. London: Hamish Hamilton.

90. Skjulstad, S., and Morrison, A. (2016). Fashion film and genre ecology. *Journal of Media Innovations*, **3**(2). Available at http://www.journals.uio.no/index.php/TJMI/issue/view/320

91. Slaughter, R. (2004). *Futures Beyond Dystopia: Creating Social Foresight*. London: Routledge/ Farmer.

92. Sloterdijk, P. (1988). *Critique of Cynical Reason*. Minneapolis: University of Minnesota Press.

93. Srnicek, N., and Williams, A. (2016). *Inventing the Future: Postcapitalism and a World Without Work*. London: Verso.

94. Stengers, I. (2011a). *Thinking with Whitehead: A Free and Wild Creation of Concepts*, Translated by M. Chase. Cambridge: Harvard University Press.

95. Stengers, I. (2011b). *Cosmopolitics II*, Translated by R. Bononno. Minneapolis: University of Minnesota Press.

96. Stuedahl, D., and Smørdal, O. (2015). Matters of becoming. Translations and enactments of social media in museums experimental zones. *Co-Design*, **11**(3), 193–207.

97. Townsend, A. (2013). *Smart Cities: Big Data, Civic Hackers, and the Quest for a New Utopia*. New York: W. W. Norton.

98. Urry, J. (2016). *What is the Future?* Cambridge: Polity Press.

99. van der Duin, P. (2016). *Foresight in Organizations: Methods and Tools*. London: Routledge.

100. van der Duin, P., Heger, T., and Schlesinger, M. (2014). Toward networked foresight? Exploring the use of futures research in innovation networks. *Futures*, **59**, 62–78.

101. Voros, J. (2003). A generic foresight process framework. *Foresight*, **5**(3), 10–21.

102. Voros, J. (2005). A generalised "layered methodology" framework. *Foresight*, **7**, 28–40.

103. Willey, A. (2016). A world of materialisms: postcolonial feminist science studies and the new natural. *Science, Technology, & Human Values*, **41**(6), 991–1014.

104. Wilson, C. (2016). *Come and Take It: The Gun Printer's Guide to Thinking Free*. New York: Gallery Books (Kindle edition).

105. Yelavich, S., and Adams, B. (2014). *Design as Future Making*. London: Bloomsbury.

Index

3D printer designer, 13
3D printer developers, 35
3D printer inventors, 25, 30–32,
    35, 45, 48–49
3D printers, 3, 5–6, 9–10, 13–14,
    22, 26–27, 30–35, 39–42,
    44–53, 57–58, 65–66, 102,
    115, 117, 126–28
3D printing, 1–4, 6–7, 13–15,
    17–18, 21–22, 24–36, 39–41,
    44–55, 57–58, 79–87, 97–99,
    102–6, 113–19, 134–35,
    137–38
3D printing technologies, 16, 26,
    40, 43–44, 57–58, 116, 134,
    144, 211
3D printing tools, 44–45, 48, 124

ABS, *see* acrylonitrile-butadiene
    styrene
abstraction, 91, 146, 148, 169–70
acrylonitrile-butadiene styrene
    (ABS), 34, 61, 63, 67
action
    bend-and-weld, 150
    negotiating, 95–96
adapt, integrate, compensate, and
    elongate (AICE), 43, 80, 93,
    100
adaptation, 39, 209
additive character, 144
additive digital fabrication
    technologies, 28
additive fabrication, 22
additive-manufactured products,
    191
additive-manufactured prototypes,
    166

additive manufacturing (AM),
    15, 18, 22, 55, 60, 78–79, 81,
    101–2, 113, 116, 137, 143,
    165, 167, 177
additive manufacturing
    technologies, 18, 55, 162
additive material production, 22
additive processes, 52, 54–55
additive product development, 137
additive shaping, 155
aesthetical approach, 145
aesthetical attributes, 122
aesthetical diversity, 147
aesthetical models, 33, 126
aesthetical qualities, 121
aesthetic development, 146
aesthetic reasoning, 145
aesthetics, 42, 132, 144–46, 201
    baroque, 208
    emergent, 161
    negotiated, 149
affordances, 26, 144, 151, 222
    technological, 27, 29
A-frame, 184–87
AICE, *see* adapt, integrate,
    compensate, and elongate
AICE model, 98, 100, 104, 106, 109
AM, *see* additive manufacturing
anteversionheads, 76–77
articulations, 111, 203, 206, 212,
    214, 220
    complex, 120
    creative, 220
    essayistic, 204
artifacts, 29–30, 47, 50–51, 54,
    58, 65, 67–68, 112–14, 116,
    119–21, 135–37, 200, 206–7,
    210–11, 213–14

3D-printed, 146
colored, 57
end-use, 41
epistemic, 215
hybrid, 120, 136
laser-sintered, 39
assemblages, 27, 120, 204

backcasting, 87
baroque, 201, 209, 221–22
Bene's experiment, 176
B-frame, 184, 186–87
binder, 57–58
boundaries, elastic, 81
brand building, 93, 165, 167, 169,
    173, 195
branding, 18, 39, 108, 165, 195–96

CAD, *see* computer-aided design
Capjon's model, 130
CATIA, 32, 104
Christmas tree model, 86
clay, 26, 105, 121, 149, 151–52,
    155, 157, 166
clay-based liquid deposition
    modeling, 91
clay models, 146, 153–54
clay printer, 92, 105
CLIP, *see* continuous layer interface
    production
CNC, *see* computer numerical
    control
CNC mills, 69–70, 132
codesign, 92–94, 107, 147
codesigning, 93, 193
coecologies, 205
color-dyeing, 119
complexity, 76, 84, 106, 121–22,
    148, 156, 166–67, 171, 202,
    212, 219, 221
composition, structural, 154
computer-aided design (CAD), 10,
    13, 32, 53, 93–95, 116, 151
computer algorithms, 99, 159

computer numerical control (CNC),
    40, 54, 104, 119
conceptualization, 18, 29, 40–41,
    125, 215
conceptualizing, 23, 31, 40–41, 48,
    86, 124, 147
continuous layer interface
    production (CLIP), 63
creativity, 120, 193, 217
critical making, 113
curing, 33, 52, 60, 63
customizations, 108–9, 167, 193,
    195, 202
customizing, 109
cybernetics, 217

DAFC, *see* diversity abstraction
    forced choosing
deconstructing, 90–91
deconstruction, 146, 152, 154, 163
deformations, 56, 155–56
deposition, rapid plasma, 65, 69
design
    computational, 42
    computer-aided, 10, 32, 93, 116,
        151
    consumer-oriented, 24
    custom, 42, 135
    digital, 144
    discursive, 207, 220–21
    engineering, 135
    engineering-driven, 123
    human-centered, 42
    parametric, 177
    user-centered, 89, 123, 138
design artifacts, 146
design-based analytical models,
    144
design-based research, 201
design development, 116, 124,
    136–38
design elements, 100, 108, 168,
    171–72, 175, 177, 186, 188,
    194–95

Design elements, User experience, and key Drivers (DUD), 175, 177–78, 186
design futures, 201, 204, 216
design methods, 82–83, 87, 93
design perspective, 18, 40, 219
design practice, 40, 44, 90, 114, 121, 123–24, 200–201, 214
  consumer-oriented, 48
  contemporary, 121
  developmental, 44
  professional, 145
  transdisciplinary, 200
design process, 13, 77–78, 80–81, 83–86, 89–90, 92–93, 97–98, 101, 107, 109, 122–25, 130–31, 153, 156, 173–74
design projects, 38, 84, 89, 130–31, 135, 149, 156, 213
design prototyping, 114–15
design representations, 41, 44, 48, 114, 122, 130–31, 134, 136
  physical, 125
  visual, 35, 122
design space, 22, 112, 114, 118, 189, 195
  digital, 137
design work, 17, 41, 146–47, 202
desktop 3D printers, 26–27, 34, 40, 45, 51, 53, 65
desktop 3D printing, 45, 134, 211
desktop fabrication, 45, 204, 228
desktop fabricators, 26, 135
desktop factory printer, 134
desktop inkjet printer, 68
desktop printers, 41, 47, 67, 211
desktop publishing, 50
desktop SLA printer, 64
developmental design tool, 31, 48
developmental methodologies, 22
developmental process, 41–42, 114, 124
developmental prototyping tools, 41

development cost, 16–17, 106
D-frame, 184, 186–87, 189
digital blueprint, 32, 47, 53, 113, 137
digital curve tool, 151
digital design tools, 9, 144
digital fabrication, 21–24, 27, 29, 44, 46, 49, 53, 113, 115, 117–18, 123–24, 137, 143, 200–201, 221
  consumption-oriented, 22
digital fabrication technologies, 43, 116, 119, 144
digital information, 21, 23, 44, 118, 120
digitalization, 42, 145, 212
digital light processing (DLP), 63
digital models, 1–2, 116, 156
digital tools, 93, 147, 151–52, 202
direct metal laser sintering (DMLS), 59
discursive design approach, 218
discursive design framing, 207
diversity abstraction forced choosing (DAFC), 91–93, 95
DIY kit, 34–35
DLP, *see* digital light processing
DLP light source, single, 64
DLP projectors, 63–64
DMLS, *see* direct metal laser sintering
double-diamond model, 85
dual-nozzle setup, 67
DUD, *see* Design elements, User experience, and key Drivers
DUD analysis, 178, 186
DUD framework, 175, 177, 186

economies-of-one, 30, 37–39, 48
economies-of-scale, 36–37
elasticity, 108, 161, 176
elastomers, 62–63
end-user parts, 55, 58

EoF, *see* evolution of form
EoF model, 149–50, 153–56, 158,
 160
ergonomics, 22, 132, 134, 170, 175
evolution of form (EoF), 147,
 150–51, 161
exploration, 86, 113, 145, 147, 161,
 175, 181, 183, 193, 220
 design-oriented, 147
 systematic abstract, 154
extruder head, 65–66

fabrication process, 22, 30, 49,
 55–56, 58, 60, 63–64, 67
 digital, 51, 53, 120
fabrication tools, 41, 49
 digital, 45–48, 50, 54, 114, 116,
 157
FDM, *see* fused deposition
 modeling
FDM/FFF, 66–67
FFF, *see* fused filament fabrication
filament deposition modeling,
 51–52, 63
fit, ergonomic, 133–34
fixtures, 44, 48, 103, 105
force
 direct, 155
 internal structural, 155
 market-driven persuasive, 218
force output, 132
forecasts, 204
 economic, 36
foresight, 201, 215–19
 strategic, 216
form
 abstract, 149, 151
 archetypal, 127
 digital, 136
 edgy, 157
 free, 161
 geometrical, 157
 organic, 150
 physical, 15, 145, 147

 virtual, 151–52
 visual, 144–46, 153, 161
formalistic language, 145
formats, 50, 83, 87, 98, 102, 156,
 207
 digital, 14, 94
 physical, 91
format wars, 25
form blowing, 156
form creation, 151
form development, 98
 radical, 161
 visual, 144
form elasticity, 156
form experiments, 149
form exploration, 149
form expressions, 153
form families, 160–61
form finding, 146, 161
 visual, 133, 161
form freedom, 165–67, 173, 189,
 194
forming methods, 65
forming principles, 69
forming process, 52
formlessness, 162
fragility, 58, 153
functionality, 90, 98–99, 104,
 125–29, 132
 improved, 117
 marketable, 8
functionality fit, 127
fused deposition modeling (FDM),
 5–6, 14, 32, 34, 51–52, 63, 65,
 77, 126
fused filament fabrication (FFF),
 65–66

Gartner's hype cycle, 4, 7, 17
generic models, 83
gigamapping, 114

hardware, 5–6, 35, 132, 151
 off-the-shelf, 35

HCD, *see* human-centered design
HCD development, 43
HCI, *see* human–computer interaction
holistic business models, 183
human-centered design (HCD), 42–43
human–computer interaction (HCI), 23, 42, 44, 72
hype cycle, 3
  generic, 3

ideo-pleasure, 172–73
industrial designers, 121–23, 126, 147, 161
industrial processes, 26–27
Industrial Revolution, 14, 36–37, 211
in-house FDM, 127
injection molding, 12, 16–17, 22, 55, 76–78, 101, 105, 117, 132–33, 150
inkjet printer, 50, 57, 62
inkjet printing, traditional, 49
innovation, 26, 43, 50, 123, 158, 170, 181, 201, 212, 217, 219, 221
  networks of, 217
  open-source, 21
  radical, 43, 72
  technological, 43
interfaces, 120, 136
  digital, 120, 221
  haptic, 151
  visual, 151
Internet of Things (IoT), 120, 202, 217
intervention, social, 23, 25
IoT, *see* Internet of Things
IoT product hybrids, 120

jigs, 13, 78, 103, 105
joints, 76–77, 80
  3D-printed, 15, 75, 80, 189

laminated object manufacturing (LOM)
laser sintering, 179
  direct metal, 59
  selective, 1, 29, 51, 76–77, 133, 179
layered methodology, generalized, 216
layers, 15, 22, 25–26, 33, 52, 55, 57, 59–62, 64–69, 81, 98, 100, 104, 106, 109
  oxygen-permeable, 63
  thin, 51, 60
  weak, 80
layer thicknesses, common, 67
layerwise fashion, 52, 57, 59, 63, 65, 67–69
LDM, *see* liquid deposition modeling
LED, *see* light-emitting diode
light-emitting diode (LED), 189
liquid
  photopolymer, 63
  photopolymeric, 51, 61–62
liquid-based processes, 52, 62
liquid deposition modeling (LDM), 91
liquid photopolymer, 60, 63
LOM (laminated object manufacturing), 65, 68–69
low-cost FDM printers, 51
low-cost FFF printers, 67
low-cost manufacturing facilities, 9
low-priced printers, 14

machines, 1–2, 4–5, 32, 58, 60, 69, 102, 117, 184
  low-cost FDM, 7
  powder-based, 52
  rapid-prototyping, 117
  sewing, 50
  vacuum-forming, 40
machine tools, new, 36
Makerbot, 5, 34–35, 45, 49

MakerBot printer, 34
maker festivals, 46
maker movement, 45, 47, 71
makers, 9, 21, 25, 27, 31, 35,
    44–46, 48–49, 85, 149, 211
maker spaces, 35, 45, 49, 143, 211
    local, 29, 46
making processes, 23, 47, 53–54,
    115–16
    natural, 54
    technological, 54
making tools, 48, 116
malleability, 144, 202
mass customization, 37, 107, 116
mass-produced artifacts, 116,
    131–32, 135
mass production, 37, 105, 125,
    129, 132, 202, 211
material artifacts, 26, 136
materialities, 52, 128, 136, 220
    analog, 120, 201
    technological, 44
materials, 31, 51–52, 54–58, 60,
    62–63, 66–67, 69–70, 104,
    118, 120, 173–74, 180–81,
    200–203, 207, 213–14
    base, 61
    binder, 57
    blended, 67
    common, 56
    digital, 98
    elastic, 155
    elastomeric, 56
    liquid-based, 61
    medical, 101
    organic, 54, 133
    photopolymeric, 62, 64
    physical, 23, 119–20, 138, 153
    plaster, 58
    plaster-based, 57
    powder-based, 61
    proprietary, 65
    prototyping, 63
    raw, 103

    sturdy, 192
material stiffness, increased, 56
MAYA principle, 158–59
metal powders, 52, 59
metals
    powdered, 59
    specialist, 69
methods
    analytical, 89
    codesign, 95
    developmental, 148
    double-diamond, 96
    user-centered, 92
mock-ups, 57, 84, 113, 122, 126,
    130–31
models, 1–3, 18, 57–59, 68, 79,
    81–88, 92–93, 106–7, 111–14,
    121–23, 125–31, 133–37, 144,
    149–50, 161
    adapted, 96
    appearance, 122
    conceptual, 132
    core, 146
    foundational, 147
    multitype, 88
model TR-55, 7
molds, 32–33, 40–41, 44, 48, 54,
    79, 120
    3D-printed, 40
    internal, 54
    silicon, 76
    silicone, 118, 128
multimaterial artifacts, 22
multitypes, 84, 130–32, 136–37
multityping, 132, 135
    rapid, 129
Mykita, 38–40, 167, 177, 180–84,
    190, 192–94, 196
Mykita brand, 179, 182–83, 191
Mykita Haus, 182–83, 191
Mykita Mylon, 177–82, 192

near-net shape, 69–70
negotiotypes, 84–88, 95, 132

incremental, 130
negotiotyping, 130, 136
new product development (NPD),
    17, 123–25, 138, 141, 203, 217
nozzle, 51, 66–67
NPD, *see* new product development

onion model, 81
overhangs, 59–61, 64, 105
    angled, 61, 64
    complex, 13
    rocky, 210

PA, *see* polyamide
PA-6, 56
PA-11, 56, 133
PA-12, 56
painting, 10, 103, 148
    abstract, 148
    spray, 56
paradigm, 2, 18, 28, 30, 36, 48,
    132, 135
    3D-printed, 82
    consumptive, 28
    long-tail production, 14
    mass-manufacturing, 22
paradox, 99, 106, 158
parametric modeling, 182
PCB, *see* printed circuit board, 119
PD, *see* product design
PET, *see* polyethylene terephthalate
physical artifacts, 23, 50, 93, 113,
    137, 144, 203
physical computing, 204, 228
physical model, 1, 122
physical products, developing, 93
physio-pleasure, 172
PLA, *see* polylactic acid
plasma arc heating, 69
plastic artifacts, 27, 37
plastic injection, 32, 125
plastic powder, 5, 101
plastic printers, 92, 105
polyamide (PA), 56, 61, 76

polyethylene terephthalate (PET),
    40, 56, 67
PolyJet, 51, 62–63
PolyJet printer, 62–63
polylactic acid (PLA), 34, 41, 61, 67
polyurethane (PUR), 125, 128–29,
    131
polyvocality, 208
postprocessing, 56, 59, 61, 64, 69,
    78–79, 98, 101
postprocessing chamber, 63
powder
    abrasive glass, 56
    fine-grained metal, 59
    fine-grained plaster, 57
    fine-powdered polyamide, 39
    gypsum, 57
    melted metal, 59
    nonsintered, 56
powder-based processes, 52, 56,
    59
pq, 167, 184–91, 193–94, 196
precision, 165, 202
precision craftsmanship, 183
printed circuit board (PCB), 119
printers
    clay, 92, 105
    industrial-grade, 67
    private, 107
product conceptualization,
    113–14, 137
product design (PD), 41–43,
    112–13, 116, 118, 120,
    136–37, 144–46, 158, 161,
    200–203, 205–7, 209, 215,
    217, 220–21
product development paradigm,
    17, 81
production
    artifact, 202
    freeform, 156
    small-scale, 149, 167, 193
production cost, 8, 16, 76, 103, 106

production methods, 78, 104,
  106–7, 134, 150, 152, 156,
  173, 192, 214
  classic, 105
  real-life, 152
production technology, 18, 158
  contemporary, 150
  modern, 38
products
  3D-printed, 105, 107–8
  beta, 128–29
  custom-made, 156
  digital, 14
  digital/physical, 136
  fast-changing, 159
  functional, 132
  heritage, 102
  intuitive, 8
  limited-edition, 38
  niche, 109
  popular artisan, 109
prototypes, 3–4, 13–14, 22–23,
  55, 82, 84–85, 92, 94, 111–13,
  119, 121, 123, 125, 127, 130
  functional, 33, 84, 86, 128
  mechanical, 58
prototyping, 34, 53, 78, 112–13,
  119, 123, 130, 132–33,
  135–36, 143–46
psycho-pleasure, 172
PUR, *see* polyurethane

QFD, *see* quality function
  deployment
quality function deployment
  (QFD), 92
qualitative manifestation, 112
quality
  near-injection-molded, 61, 64
  visual, 153
quality assurance, 27
quality control, 119

rapid manufacturing, 77

rapid multityping models, 130
rapid plasma deposition (RPD), 65,
  69–70
rapid prototyping (RP), 3–4, 6–7,
  13–15, 22, 33, 35, 40–41, 60,
  62, 76, 113, 116, 131–34, 137,
  143, 145–46
rapid tooling, 12, 129
reflection-in-action, 115
reflection-on-action, 115
releasetype, 85, 90, 95, 120, 136,
  138
  3D-printed, 104
RepRap convention, 35
RepRap movement, 117
RepRap project, 34, 66
robots, 29, 36, 212, 224
RP, *see* rapid prototyping
RPD, *see* rapid plasma deposition

Schrödinger's cat, 1–2
science, technology, and society
  (STS), 23, 50, 203, 207, 209,
  217
selective laser melting (SLM),
  58–60
selective laser sintering (SLS), 1,
  5–6, 29, 39–40, 51, 55–56, 60,
  76–77, 101, 119, 133, 179, 187
self-copying, 117
seriotypes, 84–85, 130, 132, 136
seriotyping, 130, 135–36
SLA, *see* stereolithography
SLA-DLP, *see* stereolithography
  digital light processing
SLM, *see* selective laser melting
SLS, *see* selective laser sintering
smart cities, 213
smarter homes, 213
smartphones, 8, 53
smart watches, 120
socio-pleasure, 172
software, 35–36, 44, 119, 134, 146,
  151–52, 158, 161, 205, 211

solid-based processes, 65, 67
stages
  conceptual, 134
  creative, 78
  engineering, 104
  fuzzy, 124
  implementation, 173
  value-making, 29
  visiotype, 84
stereolithography (SLA), 4–6, 29,
    33, 51–52, 60–64, 76–77, 116,
    140, 156
stereolithography digital light
    processing (SLA-DLP), 63–64
STS, *see* science, technology, and
    society
subtractive material production,
    22
subtractive shaping, 155
SunBell, 114, 124–29, 131, 134–35,
    137
support scaffolds, 105
support structures, 59–61, 64, 70,
    105
  soluble, 63
sustainability, 115

taxonomy, 122, 130
technodeterminist logic, 205
technodeterminists, 203, 214–15
technological artifacts, 23, 25,
    30–31, 36
technological development, 22–23,
    26, 49–50, 202
  nonlinear, 24
technologies, 2–8, 12–15, 25–29,
    43–44, 52–55, 68–70, 76–77,
    87–88, 100–101, 143–44,
    165–66, 179–81, 190–93,
    200–201, 203–4
  advanced, 196
  digital, 177, 212
  emergent, 211
  input, 151–52

low-cost, 53
model-making, 4
multijetting, 62
transformative, 113
tessellations, 106, 109
thermoplastic, 65
  common, 67
  organic, 67
titanium, 59–60, 65, 69–70
  aerospace-grade, 69
tools
  computational, 211
  democratized, 49
  developmental, 21, 40
  engineering-oriented, 35, 48
  ergonomic, 170
  molding, 128
  preassembled, 45
  predefined, 54
  prototyping, 33, 68, 113, 137
  simulation, 98, 104
  surface-based, 166
track
  hermeneutical, 28
  technological, 15, 28
transdisciplinary analysis, 202

user-centered approach, 124
user customization, 38
user experience, 123, 172, 175,
    177, 186, 188, 196
  actual, 172
  intended, 171, 194–95
  new, 196
user function, 153
user functionality, 132
user interface, 136, 162
UV laser, 60
UV laser cures, 60
UV light, 63
  intensive, 61

value creation, 29, 35, 44

VCR, *see* video cassette recorder
VDR, *see* visual design
    representation
vehicles, moving 3D-printing, 102
VHS, *see* Video Home System
video cassette recorder (VCR),
    25–26, 30
Video Home System (VHS), 5, 25
video technology, 25–26
virtual reality (VR), 42
visiotypes, 84, 86–87, 92, 130, 132
visiotyping, 130, 136
visual cues, 171

visual design cues, 197
visual design representation
    (VDR), 35, 48, 122, 130, 140
visualizations, 78, 82, 85–86, 88,
    92, 108, 151
VR, *see* virtual reality

web-based self-services, 120, 139
woodblock printing, 50

zerotypes, 84, 86–88, 92
zerotypes/visiotypes, 92
ZPrinters, 57